AF801783

Askar A. Tuganbaev

Arithmetical Rings and Endomorphisms

Also of Interest

Modules over Discrete Valuation Rings, 2nd edition
Piotr A. Krylov/Piotr A. Tuganbaev, 2018
ISBN 978-3-11-060977-6, e-ISBN (PDF) 978-3-11-061114-4,
e-ISBN (EPUB) 978-3-11-060985-1

Approximations and Endomorphism Algebras of Modules
Volume 1 – Approximations/Volume 2 – Predictions
Rüdiger Göbel/Jan Trlifaj, 2012
ISBN 978-3-11-021810-7, e-ISBN (PDF) 978-3-11-021811-4

Progress in Commutative Algebra 1
Combinatorics and Homology
Ed. by Francisco, Christopher/Klingler, Lee C./Sather-Wagstaff, Sean/Vassilev, Janet C., 2012
ISBN 978-3-11-025034-3, e-ISBN (PDF) 978-3-11-025040-4

Progress in Commutative Algebra 2
Closures, Finiteness and Factorization
Ed. by Francisco, Christopher/Klingler, Lee C./Sather-Wagstaff, Sean M./Vassilev, Janet C., 2012
ISBN 978-3-11-027859-0, e-ISBN (PDF) 978-3-11-027860-6

Algebra and Its Applications
Proceedings of the International Conference held at Aligarh Muslim University, 2016
Ed. by Ashraf, Mohammad/De Filippis, Vincenzo/Rizvi, Syed Tariq, 2018
ISBN 978-3-11-054092-5, e-ISBN (PDF) 978-3-11-054240-0,
e-ISBN (EPUB) 978-3-11-054098-7

Askar A. Tuganbaev

Arithmetical Rings and Endomorphisms

DE GRUYTER

Mathematics Subject Classification 2010
16D25, 16D40, 13F05, 16D50, 16D70, 16E60, 16P40, 16P50, 16P60, 13F05

Author
Prof. Dr. Askar A. Tuganbaev
National Research University MPEI
Department of Higher Mathematics
Krasnokazarmennaya 14
Moscow 111250
Russian Federation
tuganbaev@gmail.com

ISBN 978-3-11-065889-7
e-ISBN (PDF) 978-3-11-065982-5
e-ISBN (EPUB) 978-3-11-065915-3

Library of Congress Control Number: 2019939556

Bibliographic information published by the Deutsche Nationalbibliothek
The Deutsche Nationalbibliothek lists this publication in the Deutsche Nationalbibliografie; detailed bibliographic data are available on the Internet at http://dnb.dnb.de.

© 2019 Walter de Gruyter GmbH, Berlin/Boston
Cover image: jjaakk / DigitalVision Vectors / Getty Images
Typesetting: le-tex publishing services GmbH, Leipzig
Printing and binding: CPI books GmbH, Leck

www.degruyter.com

Preface

This monograph is a comprehensive account of not necessarily commutative arithmetical rings, examining structural and homological properties of modules over arithmetical rings and summarizing the interplay between arithmetical rings and other rings. Modules with extension properties of submodule endomorphisms are also systematically studied in this book.

Graduate students and researchers interested in ring theory and module theory will find this book particularly valuable. Containing numerous examples, Arithmetical Rings and Endomorphisms is a largely self-contained and accessible introduction to the topic, assuming a solid understanding of basic algebra.

The study is supported by the Russian Science Foundation, project no. 16-11-10013.

Key words: arithmetical ring, distributive module, flat module, localization by a maximal ideal, Bezout ring, Hermite ring, endomorphism-extendable module, automorphism-extendable module, automorphism-invariant module, injective module, quasi-injective module, strongly semiprime ring

https://doi.org/10.1515/9783110659825-201

Introduction

This book consists of two parts. In Part I, "Arithmetical rings," we systematically study not necessarily commutative rings with distributive lattice of two-sided ideals. In Part II, "Extension of automorphisms and endomorphisms," we study modules with the extension property of automorphisms and endomorphisms from submodules to the whole module, and also characteristic submodules of their injective hulls.

The main content of the book

The main results of Section 1 "Saturated Ideals and Localizations" are Theorems 1A, 1B and 1C.

1A Theorem (Tuganbaev [172]). A right invariant ring A is arithmetical if and only if for its maximal ideal M, all $\{A \setminus M\}$-saturated ideals of the ring A form a chain with respect to inclusion.

1B Theorem (Jensen [98]). A commutative ring A is arithmetical if and only if for its maximal ideal M, the localization A_M is a uniserial ring.

1C Theorem (Tuganbaev [157]). A right invariant ring A is an arithmetical semiprime ring if and only if for its maximal ideal M, the right localization A_M exists and is a right uniserial domain.

The main results of Section 2, "Finitely Generated Modules and Diagonalizability," are Theorems 2A and 2B.

2A Theorem (Golod [77]). If A is a commutative ring, then A is arithmetical if and only if $B + r(X) = r(X/XB)$ for every finitely generated A-module X and each ideal B of the ring A.

2B Theorem (Tuganbaev [187]). If A is a right invariant, diagonalizable[1] ring, then $B + r(X) = r(X/XB)$ for every finitely generated right A-module X and each ideal B of the ring A.

The main results of Section 3 "Rings with flat and quasiprojective ideals" are the following Theorems 3A, 3B and 3C.

3A Theorem (Tuganbaev [157, 161, 182]). For an invariant semiprime ring A, the following conditions are equivalent.

1 The definition of a diagonalizable ring is given in 2.2.1.

https://doi.org/10.1515/9783110659825-202

1) A is an arithmetical ring.

2) Every submodule of any flat A-module is a flat module.

3) Every finitely generated ideal of the ring A is a quasiprojective right A-module.

3B Theorem (Jensen [98]). A commutative ring A is an arithmetical semiprime ring if and only if every submodule of any flat A-module is a flat module.

3C Theorem (Tuganbaev [182]). If A is an invariant ring, then A is an arithmetical ring if and only if every one of its finitely generated ideals is a quasiprojective right A-module such that all endomorphisms can be extended to endomorphisms of the module A_A.

The main results of Section 4 "Hermite rings and Pierce stalks" are Theorems 4A and 4B.

4A Theorem (Tuganbaev [183]). If A is a right PP Bezout ring without noncentral idempotents, then A is a Hermite ring.

4B Theorem (Tuganbaev [183]). If A is a Bezout ring such that every Pierce stalk is a serial ring, then A is a diagonalizable ring.

The main results of Section 5 "Bezout Rings, Krull dimension" are Theorems 5A, 5B and 5C.

5A Theorem (Tuganbaev [183]). If A is a Bezout exchange ring without noncentral idempotents, then A is a diagonalizable ring.

5B Theorem (Tuganbaev [187]). If A is a right invariant, a right Bezout, exchange ring, then $B + r(X) = r(X/XB)$ for every finitely generated right A-module X and each ideal B of the ring A.

5C Theorem (Tuganbaev [188]). If A is a commutative arithmetical ring, then A has the Krull dimension if and only if every factor ring of the ring A is finite-dimensional and does not have idempotent proper essential ideals.

The main results of Section 6 "Semi-Artinian and Nonsingular Modules" are Theorems 6A, 6B and 6C.

6A Theorem (Tuganbaev [184]). If M is a semi-Artinian[2] module, then M is an automorphism-extendable module if and only if M is an automorphism-invariant module.

6B Theorem (Tuganbaev [174]). If M is a module over an Artinian serial ring, then M is an automorphism-extendable module if and only if M is a quasi-injective module.

2 A module M is said to be *semi-Artinian* if each of its nonzero factor modules has a simple submodule.

6C Theorem (Tuganbaev [176]). Let $M = T \oplus U$, where T is an injective module, U is a nonsingular module, and $\mathrm{Hom}(T', U) = 0$ for any submodule T' of the module T. The module M is automorphism-extendable if and only if U is an automorphism-extendable module.

The main results of Section 7 "Modules over Strongly Prime and Strongly Semiprime Rings" are Theorems 7A and 7B.

7A Theorem (Tuganbaev [179]). If A is a right strongly prime ring, then a right A-module M is automorphism-invariant if and only if either M is a singular automorphism-invariant module or M is an injective module.

7B Theorem (Tuganbaev [176]). If M is a right module over an invariant hereditary domain A, then the following conditions are equivalent.

1) M is an automorphism-extendable (strongly automorphism-extendable) module.
2) M is an endomorphism-extendable (strongly endomorphism-extendable) module.
3) Either M is a quasi-injective singular module or M is an injective module that is not singular, or $M = X \oplus Y$, where X is an injective singular module and the module Y is isomorphic to a nonzero submodule in Q_A, where Q is a division ring of fractions of the domain A.

The main results of Section 8 "Endomorphism-extendable Modules and Rings" are Theorems 8A and 8B.

8A Theorem (Tuganbaev [167]). A ring A is a right endomorphism-extendable, right nonsingular ring if and only if $A = B \times C$, where B is a right injective regular ring, C is a left invariant, reduced Baer ring and C is a right completely integrally closed subring of its maximal right rings of fractions Q.

8B Theorem (Tuganbaev [162]). A ring A is a right (left) Noetherian ring such that all cyclic right (left) modules are endomorphism-extendable if and only if $A = A_1 \times \cdots \times A_n$, where A_i is either a simple Artinian ring or a uniserial Artinian ring, or an invariant hereditary Noetherian domain, $i = 1, \ldots, n$.

The main results of Section 9 "Automorphism-invariant nonsingular modules and the rings" are Theorems 9A, 9B, 9C, 9D.

9A Theorem (Tuganbaev [185]). Let A be a right strongly semiprime ring. If X is a right A-module and there exists an essential right ideal B of the ring A such that X is injective with respect to the module B_A, then X is an injective module.

9B Theorem (Tuganbaev [189]). If A is a ring with right Goldie radical G, then the following conditions are equivalent.

1) Every nonsingular right A-module X, which is injective with respect to some essential right ideal of the ring A, is an injective module.
2) A/G is a right strongly semiprime ring.

9C Theorem (Tuganbaev [189]). For a ring A with right Goldie radical G, the following conditions are equivalent.
1) A/G is a semiprime right Goldie ring.
2) Any direct sum of automorphism-invariant nonsingular right A-modules is an automorphism-invariant module.
3) Any direct sum of automorphism-invariant nonsingular right A-modules is an injective module.

9D Theorem (Tuganbaev [186]). A ring A is a right automorphism-invariant, right nonsingular ring if and only if $A = S \times T$, where S is a right injective regular ring and T is a strongly regular ring which contains all invertible elements of its maximal right ring of fractions.

The proof of the above Theorems 1A–9D is decomposed into several assertions, some of which are of independent interest.

All rings are assumed to be associative with a nonzero identity element. Modules are assumed to be unitary and, unless otherwise specified, all modules are right modules. A "Bezout ring" means a "right and left Bezout ring."

Distributive lattices. Arithmetical modules and rings. Distributive modules and rings

The main purpose of Part I of this book is the study of not necessarily commutative arithmetical rings.

a. A lattice $\mathcal{L}$ with operations $\cap$ and $+$ is said to be *distributive* if the following two equivalent[3] conditions hold.
i) $X \cap (Y + Z) = X \cap Y + X \cap Z$ for all $X, Y, Z \in \mathcal{L}$;
ii) $(X + Y) \cap (X + Z) = X + Y \cap Z$ for all $X, Y, Z \in \mathcal{L}$.

b. A ring with a distributive lattice of ideals is called an *arithmetical* ring.
It is directly verified that a ring A is arithmetical if and only if $X \cap (Y + Z) = X \cap Y + X \cap Z$ for all 1-generated ideals X, Y, Z of A.

c. For a module X, a submodule M of X is called a *fully invariant* (resp., characteristic) submodule in X if $\alpha(M) \subseteq M$ for every endomorphism (resp., automorphism) α of the module X.
A module M is said to be *arithmetical* if lattice $\mathcal{L}$ of all its fully invariant submodules is distributive.

3 The equivalence of conditions a) and b) is well known; e.g., see [83, Section 1.4, Lemma 10].

Since the ideals of the ring A coincide with the fully invariant submodules in A_A and with fully invariant submodules in ${}_AA$, a ring A is arithmetical if and only if the module A_A is arithmetical, if and only if the module ${}_AA$ is arithmetical.

d. A module is said to be *distributive* if the lattice of all its submodules is distributive. A module is said to be *uniserial* if any two of its (cyclic) submodules are comparable with respect to inclusion.
Every uniserial module is distributive and every distributive module is arithmetical. The ring of integers $\mathbb{Z}$ is a distributive nonuniserial $\mathbb{Z}$-module. The direct sum of $\mathbb{Z}/2\mathbb{Z} \oplus \mathbb{Z}/2\mathbb{Z}$ is an arithmetical nondistributive $\mathbb{Z}$-module.

e. Let A be a simple ring that is not a division ring (for example, let A be the ring of all 2×2 of matrices over a division ring). Then A is an arithmetical ring that is not right or left distributive.

f. A module M is said to be an *invariant* or *duo* module, if all its submodules are fully invariant in M.
A ring is said to be *right* (resp., *left*) *invariant* if all its right (resp., left) ideals are ideals, i.e., A is a right invariant (resp. left) A-module. A ring A is left (resp., right) invariant if and only if ${}_AA$ (resp., A_A) is an invariant module.
It is clear that a right invariant ring is arithmetical if and only if it is right distributive.
Left and right invariant rings are called *invariant* or *duo* rings.
All commutative rings are invariant. If A is a noncommutative division ring, then A and $A[[x]]$ are noncommutative invariant rings.

A remark on commutative arithmetical rings

In commutative algebra, arithmetical rings play an important role; they enter into the characterization of various interesting and important rings because:
i) commutative arithmetical rings coincide with commutative rings such that all their localizations by maximal ideals are uniserial rings [98];
ii) all submodules of flat modules over a commutative ring A are flat if and only if A is an arithmetical semiprime ring [98].
Arithmetical rings also appear when solving many other problems of commutative and homological algebra as necessary or/and sufficient conditions. For example, see [12, 28, 47, 50, 52, 53, 67, 68, 74–77, 88, 98, 99, 125, 126, 136, 180, 182, 188, 191, 192].

Bezout modules and rings

A module is called a *Bezout module* if all its finitely generated submodules are cyclic. Every uniserial module is a distributive Bezout module. The ring of integers is a commutative distributive nonuniserial Bezout ring.

a. Every right invariant, right Bezout ring A is right distributive; in particular, it is arithmetical. In this case, every cyclic right A-module is distributive, and every factor ring of the ring A is a right invariant, right distributive, right Bezout ring.
Indeed, it is sufficient to prove that $X = X \cap Y + X \cap Z$ for any principal right ideals X, Y, Z of the ring A with $X \subseteq Y + Z$. Since A_A is a Bezout module, $Y + Z$ is a cyclic right module over the right invariant ring A. Therefore, there exists an ideal B in A such that

$$X = (Y + Z)B = YB + ZB \subseteq X \cap Y + X \cap Z \subseteq X\,.$$

b. It is easy to verify that distributivity of the module is equivalent to distributivity of all its 2–generated submodules. Therefore, the distributivity of a Bezout module is equivalent to the property that each of its cyclic submodules is distributive.

c. It follows from the two previous assertions that every right Bezout module M over a right invariant ring A is a distributive module.

A remark on noncommutative arithmetical rings

In contrast to the commutative case, the class of arbitrary noncommutative arithmetic rings is too large for meaningful study, since it includes all simple rings, all (von Neumann) regular rings, all hereditary Noetherian semiprime rings, all biregular rings, and other large classes of rings (see Sections "Semidistributive and serial modules" (c) on page XIV and "Rings with idempotent ideals" (a,b) on page XV). Therefore, it is necessary to consider arithmetical rings that are quite close to commutative rings; however, in this case, the investigation also becomes much more difficult than in the commutative case.
For example, even if we consider invariant arithmetic rings with strong additional conditions, then, as the example below shows, there are no analogs of localizations with respect to maximal ideals that exist for any commutative ring, so the localizations are an important tool in the study of commutative rings.

An example of a noncommutative invariant arithmetical ring *A* without localizations by maximal ideals

See [137, 168]. In this example, the ring A is also a subdirectly indecomposable semilocal Bezout ring with exactly two maximal right (left) ideals.

Let $\mathbb{Z}$ be the ring of integers, $\mathbb{Q}$ be the field of rational numbers, and let $\mathbb{Q}[i]$ be the field of fractions of the ring of Gaussian integers $\mathbb{Z}[i]$ ($i^2 = -1$). We denote by φ the automorphism $q_1 + q_2 i \to q_1 - q_2 i$ of the field $\mathbb{Q}[i]$. Let R_1, R_2 be the localizations of the commutative principal ideal domain $\mathbb{Z}[i]$ with the prime ideals generated by prime elements $2+i$ and $2-i$, respectively. Then R_1 and R_2 is a commutative uniserial principal ideal domain. We set $R = R_1 \cap R_2$. Let X, Y be the ideals in R generated by the elements $2 + i$ and $2 - i$, respectively.
The commutative principal ideal domain R has exactly two maximal ideals X and Y, $R_X = R_1$, $R_Y = R_2$, $\varphi(R_X) = R_Y$, $X + Y = R$, $J(R) = X \cap Y = (2 + i)(2 - i)R = 5R$ and the factor ring $R/J(R)$ is isomorphic to the direct product of the fields A/X and A/Y. The ring R coincides with the set of all irreducible rational Gaussian fractions whose denominators are not divided by $2 + i$ or $2 - i$.
We denote by M the right R-module $\mathbb{Q}[i]/R_X$. Then all proper submodules in M_R are cyclic and form a properly ascending infinite countable chain $0 = s_0R \subset s_1R \subset s_2R \subset \dots$, where the simple module s_1R is isomorphic to the module $R/Y = R/(2 - i)R$ and $r_R(s_n) = (2 - i)^n R$ for all positive integers n. We also turn M into a left R-module with the rule $rm = m\varphi(r)$ for all elements $r \in R$ and $m \in M$. It is directly verified that M is an R–R–bimodule, which is uniserial Artinian[4] divisible right (left) R-module.
We denote by A the *trivial extension* of the R–R–bimodule M by the ring R. We recall that A is the external direct sum of the Abelian groups R, M with multiplication such that

$$(r_1, m_1)(r_2, m_2) = (r_1r_2, r_1m_2 + m_1r_2) = (r_1r_2, m_2\varphi(r_1) + m_1r_2)$$

for any $r_1, r_2 \in R$ and $m_1, m_2 \in M$. The pair $(1, 0)$ is the identity element of the ring A. We identify M, X and Y with ideals $(0, M)$, $(X, 0)$ and $(Y, 0)$ in A, respectively. The ring R is identified with the subring $(R, 0)$ in A.
The right (left) ideals of the ring A, contained in M, coincide with the R-submodules in M. We note that $mA = Am$ for any $m \in M$, the factor ring $A/J(A)$ is isomorphic to the direct product of the fields A/X and A/Y, $M^2 = 0$, the factor ring A/M is isomorphic to the commutative principal ideal domain R, M is a uniserial Artinian divisible right (left) A-Bezout module, simple right (left) A-module $s_1A = As_1$ is the least nonzero right (left) ideal in A, $r_A(s_1) = Y$ and $\ell_A(s_1) = X$. We list some properties of the ring A.
i) $X = (2 + i)A = \ell_A(s_1) = N_\ell$ and $Y = (2 - i)A = r_A(s_1) = N_r$.
ii) The element $(2 + i) + (2 - i) = 4 = 5 - 1$ is invertible in A, since $5 \in J(A)$.
iii) If we set $x = (2 + i)/4 \in X$, then $1 - x = (2 - i)/4 \in Y$, $(2 + i)s_1 = s_1\varphi(2 + i) = s_1(2-i) = 0$, $xs_1 = s_1(1-x) = 0$, $x \in X\setminus J(A) = X\setminus Y \subseteq N_\ell$, $1-x \in Y\setminus X = Y\setminus J(A) \subseteq N_r$, $s_1 \in r(x) \cap \ell(1 - x)$, $r(s_1) \cap (A \setminus Y) = Y \cap (A \setminus Y) = \varnothing$, $\ell(s_1) \cap (A \setminus X) = X \cap (A \setminus X) = \varnothing$.
iv) There does not exist the right-sided analog of the localization of the ring A by the maximal (right) ideal X and the left-sided analog of the localization of the ring A by the maximal (left) ideal Y.

---

4 A module is said to be *Artinian* if every properly descending chain in its submodules is finite.

v) $J(A)$ does not contains all right (left) zero-divisors of A.
vi) The ideal M is comparable with respect to inclusion with any right or left ideal of the ring A.
vii) A is a distributive ring.
viii) A is a subdirectly indecomposable, invariant Bezout ring.

Properties **i–iii** are directly verified. Properties **iv, v** follow from the property **iii**. The property **vi** follows from the property that M is a divisible R-module. The property **vii** follows from **vi** and the following properties: A/M is a commutative Bezout ring, M is an A-Bezout module, $mA = Am$ for every $m \in M$, and s_1A is the least nonzero right (left) ideal in A. The property **viii** follows from the property **vi** and the property that every right Bezout module over a right invariant ring is distributive by III(c).

Semidistributive and serial modules

Any direct sum of distributive (resp., uniserial, simple) modules is called a *semidistributive* (resp., *serial, semisimple*) module.

a. It is directly verified that all semidistributive (for example, serial) modules are arithmetical; e.g., see [109, 110, 195–198].

b. For a ring A, let the intersection of any two nonzero ideals of A be a nonzero ideal (e.g., this is the case if the ring A is prime). If every proper factor ring of A is arithmetical (e.g., this is the case if A is right semidistributive), then A is an arithmetical ring. It is sufficient to prove that $X = X \cap Y + X \cap Z$ for any of nonzero ideals X, Y, Z with $X \subseteq Y + Z$. By assumption, $X \cap Y \cap Z \neq 0$. Let $h\colon A \to A/(X \cap Y \cap Z)$ is a natural ring homomorphism. Since the ring $h(A)$ is arithmetical, $h(X) = h(X) \cap h(Y) + h(X) \cap h(Z)$. Therefore, $X + X \cap Y \cap Z = X \cap Y + X \cap Z + X \cap Y \cap Z$. Then $X = X \cap Y + X \cap Z$.

c. Every hereditary Noetherian semiprime ring A is an arithmetical ring.
Since A is a finite direct product of hereditary Noetherian prime rings (see [44] or [63, Theorem 20.30]), we can assume that A is a hereditary Noetherian prime ring. Then every proper factor ring of the ring A is serial (see [58, Corollary 3.2] or [63, Theorem 25.5.1]). By the previous assertion, A is an arithmetical ring.

d. Let A be a 5-dimensional algebra over a field F generated by all 3×3 matrices of the form

$$\begin{pmatrix} f_{11} & f_{12} & f_{13} \\ 0 & f_{22} & 0 \\ 0 & 0 & f_{33} \end{pmatrix},$$

where $f_{ij} \in F$. It is directly verified that $e_{11}A = e_{11}F + e_{12}F + e_{13}F$ is an indecomposable distributive Noetherian Artinian nonuniserial A–Bezout module and A is a right semidistributive, left serial, Artinian ring which is not right serial. In particular, A is a 5-dimensional arithmetical F-algebra which is not right serial.

Rings with idempotent ideals

a. If every 1-generated ideal of the ring A coincides with its square (in this case, any ideal coincides with its square), then A is an arithmetical ring.
Let X, Y, Z be three ideals of the ring A. Then

$$X \cap (Y + Z) = (X \cap (Y + Z))^2 \subseteq X(Y + Z) = XY + XZ \subseteq X \cap (Y + Z)\,.$$

b. Let A be either a *biregular* ring (i.e., every 1-generated ideal of the ring A is generated by a central idempotent or *regular* ring (i.e., every principal right (left) ideal is generated by idempotent). It is well known and it is directly verified that every 1-generated ideal of the ring A coincides with its square. By the previous assertion, A is an arithmetical ring.

Automorphism-invariant, pseudo-injective and quasi-injective modules

One of the principal purposes of Part II of this book is the study of automorphism-invariant modules and related modules.
A module M is said to be *automorphism-invariant* (resp., *endomorphism-invariant*), if it is a characteristic (resp., *fully invariant*) submodule of its injective hull.

In [60, Theorem 16], it is proven that a module M is an automorphism-invariant if and only if M is a *pseudo-injective* module, i.e., if for any submodule X in M, every monomorphism $X \to M$ can be extended to an endomorphism of the module M.

Pseudo-injective modules were studied in several papers; e.g., see [60, 96, 154]. Automorphism-invariant modules were studied in several papers; e.g., see [7, 56, 60, 85, 116, 148, 174, 175, 177–179, 181].

a. A module M is said to be *injective with respect to the module X* or *X-injective* if for any submodule X_1 in X every homomorphism $X_1 \to M$ can be extended to a homomorphism $X \to M$.
A module M over the ring the A is said to be *injective* if M is injective with respect to any A-module.
For example, over finite direct products of rings of matrices over division rings all modules are injective. In addition, an Abelian group M is an injective module over the ring of integers $\mathbb{Z}$ if and only if M is a divisible Abelian group, i.e., M is a direct sum of groups that are isomorphic to the additive group $\mathbb{Q}$ of rational numbers and quasi-cyclic groups $\mathbb{Z}(p^\infty)$.
Injective objects play a very important role in many mathematical categories. In this book, we consider only injective right modules over the ring A, i.e., injective objects of the category Mod-A of all right A-modules. One of the reasons for the importance of injective modules is that every module M is an essential submodule of some injective

module X, which is called the *injective hull* of the module M and the injective hull X is uniquely (up to isomorphism) defined. For example, if $\mathbb{Z}$ is a ring of integers, then the additive group $\mathbb{Q}$ is the injective hull of the module $\mathbb{Z}_{\mathbb{Z}}$. We note that $\mathbb{Z}$ is not an automorphism-invariant $\mathbb{Z}$-module, since $\alpha(\mathbb{Z}) \not\subseteq \mathbb{Z}$, where $\alpha\colon q \to q/2$ is an automorphism of the $\mathbb{Z}$-module $\mathbb{Q}$, which is the injective hull of $\mathbb{Z}_{\mathbb{Z}}$.

b. A module M is said to be *quasi-injective* or *self-injective* if M is injective with respect to itself.
It is clear that every injective module is quasi-injective. Every finite cyclic group is a quasi-injective (automorphism-invariant) noninjective module over the ring $\mathbb{Z}$.

It is well known that a module M is quasi-injective if and only if M is an endomorphism-invariant module, i.e., $\alpha(M) \subseteq M$ for any endomorphism α of the injective hull of the module M, see [102] or [194, 17.11]. This implies that every quasi-injective module is an automorphism-invariant module. The converse assertion is not true; this follows from the following example.

c. Example. Let $\{F_{i=1}^{\infty}\}$ be a countable set of copies of the field $\mathbb{Z}/2\mathbb{Z}$, R be the direct product of all the F_i, and let A be the subring of R consisting of all sequences that are stabilized at finite step. In [60, Example 9], it is proven that A is an automorphism-invariant A-module that is not quasi-injective.

Automorphism-extendable and endomorphism-extendable modules

A module M is said to be *automorphism-extendable* (resp., *endomorphism-extendable* if for any submodule X in M every automorphism (resp., endomorphism) of the module X can be extended to an endomorphism of the module M.

a. It is clear that every quasi-injective module is endomorphism-extendable and every endomorphism-extendable module is automorphism-extendable. In addition, all direct summands of automorphism-extendable (resp., endomorphism-extendable) modules are automorphism-extendable (resp., endomorphism-extendable) modules.

If M is an automorphism-invariant module, then every automorphism of any submodule of the module M can be extended to an automorphism of the module M, see 6.1.3 below; in particular, M is an automorphism-extendable module. In addition, $\mathbb{Z}$ is an example of an automorphism-extendable (endomorphism-extendable) $\mathbb{Z}$-module which is not automorphism-invariant (quasi-injective).

b. Let A be a regular ring. If A_A is an endomorphism-extendable module, it follows from Theorem 8.2.2 of this book that the module A_A is injective; also see [167].

c. The following three examples show that an automorphism-extendable module M_A is not necessarily endomorphism-extendable even if either $M = A$ is a finite-dimensional algebra over a finite field or $M = A$ is a simple principal right (left) ideal domain, or $M = A$ is a commutative regular ring.

d. Example. Let F be the field of order 2 and let A be a finite 5-dimensional algebra over field F formed by all (3×3)-matrices of the form

$$\begin{pmatrix} f_{11} & f_{12} & f_{13} \\ 0 & f_{22} & 0 \\ 0 & 0 & f_{33} \end{pmatrix},$$

where $f_{ij} \in F$. In [148], it is proven that $e_{11}A = e_{11}F + e_{12}F + e_{13}F$ is a finite cyclic automorphism-invariant projective module that is not quasi-injective. By Theorem 6.1.10, every endomorphism-extendable Artinian module is quasi-injective. Therefore, $e_{11}A$ is an automorphism-extendable module which is not endomorphism-extendable.

e. Example. Let F be a field and A an algebra over F with two generators x, y and one defining relation $xy - yx = 1$. We prove that A_A and $_AA$ are automorphism-extendable modules and the modules A_A and $_AA$ are not endomorphism-extendable. It is well known that A is a simple principal right (left) ideal domain, A is not a division ring, and the group of invertible elements $U(A)$ of the domain A coincides with $F \setminus 0$. In particular, $U(A)$ is contained in the center of the domain A. Let a be a nonzero noninvertible element of the domain A. It is sufficient to prove the following two assertions.
$(*)$ For every automorphism α of the module aA_A, there exists an invertible element u of the ring A such that $\alpha(ab) = uab$ for all $b \in A$.
$(**)$ There exists an endomorphism f of the module aA that cannot be extended to an endomorphism of the module A_A.

Since $\alpha(aA) = aA$, we have that $\alpha(a) = au$ and $a = auv$ for some elements $u, v \in A$. Then $uv = 1$. Since A is a domain, $vu = 1$ and $u \in U(A) \subset F$. Then $uab = aub = \alpha(ab)$ for all $b \in A$, and $(*)$ have been proven.

Since A is a simple domain and a is a nonzero noninvertible element, $AaA = A \neq Aa$. Therefore, $ab \not\subseteq Aa$ for some element $b \in A$. Since A is a Noetherian domain, A has the classical division ring of fractions which contains element a^{-1}. Then $aba^{-1}aA \subseteq aA$ and the relation $f(ac) = aba^{-1}c$ defines an endomorphism f of the submodule aA_A in A_A. We assume that f can be extended to an endomorphism φ of the module A_A. We set $d = \varphi(a)$. Then

$$ab = aba^{-1}a = f(a)a = \varphi(a)a = da \in Aa\,.$$

This is a contradiction; $(**)$ have been proven. □

e. Example. Let $\{F_{i=1}^{\infty}\}$ be a countable set of copies of the field $\mathbb{Z}/2\mathbb{Z}$ and let A be the subring of the direct product of all the F_i consisting of all sequences stabilized at finite step. Then A is a commutative regular ring. In [60, Example 9], it is proven that A is an automorphism-invariant module that is not quasi-injective. In particular, A_A is an automorphism-extendable module. By **b**, the automorphism-extendable module A_A is not endomorphism-extendable.

Some definitions and notations

Let A be a ring, X be a right A-module, Y be a left A-module, X_1 and X_2 be two subsets in X, Y_1 and Y_2 be two subsets in Y, $\max X$ be the set of all maximal submodules in X, and let $J(X)$ and $\operatorname{End} X$ be the Jacobson radical and the endomorphism ring of the module X, respectively.

If $f\colon X \to Q$ is a homomorphism of modules and M is a submodule in Q, then we denote by $f^{-1}(M)$(M) the submodule $\{x \in X \mid f(x) \in M$ of the module X.

a. For any subset B of the ring A, we denote by ABA the ideal in A generated by the set B.
We denote by $(X_1 \,^{\cdot}.\, X_2)$ the subset $\{a \in A \mid X_1 a \subseteq X_2\}$ of the ring A. If X_2 is a submodule in X, then $(X_1 \,^{\cdot}.\, X_2)$ is a right ideal of the ring A. If X_1, X_2 are two submodules in X, then $(X_1 \,^{\cdot}.\, X_2)$ is an ideal in A.
A subset $\{a \in A \mid aY_1 \subseteq Y_2\}$ of the ring A is denoted by $(Y_2 \,.^{\cdot}\, Y_1)$. If Y_2 is a submodule in Y_1, then $(Y_2 \,.^{\cdot}\, Y_1)$ is a left ideal of the ring A. If Y_1, Y_2 are two submodules in X, then $(Y_2 \,.^{\cdot}\, Y_1)$ is an ideal in A.

b. We denote by $r(X_1)$ the *right annihilator* $(X_1 \,^{\cdot}.\, 0) = \{a \in A \mid Xa = 0\}$ of the set X_1 which is a right ideal in A. If X_1 is a submodule in X, then $r(X_1)$ is an ideal in A.
We denote by $\ell(Y_1)$ the *left annihilator* $(0 \,.^{\cdot}\, Y_1) = \{a \in A \mid aY_1 = 0\}$ of the set Y_1 which is left ideal in A. If Y_1 is a submodule in N, then $\ell(Y_1)$ is an ideal in A.

c. A ring A is said to be *local* if for any element $a \in A$, at least one of the elements $a, 1 - a$ is invertible. A ring A is local if and only if its factor ring $A/J(A)$ modulo the Jacobson radical $J(A)$ is a division ring.
A ring A is said to be *semilocal* if its factor ring $A/J(A)$ modulo the Jacobson radical $J(A)$ is isomorphic to the finite direct product of matrix rings over division rings.
A ring A is said to be *semiperfect* if A is semilocal and all idempotents of the factor ring $A/J(A)$ can be lifted idempotents of the ring A.

d. A ring A is called an *exchange* ring if for any element $a \in A$, there exists an idempotent $e \in aA$ such that $1 - e \in (1 - a)A$. The property to be an exchange ring is equivalent to its left-sided analog; see [132] or [168].
In [132], it is proven that if A is a ring and the factor ring $A/J(A)$ is a regular ring such that all idempotents can be lifted idempotents of the ring A, then A is an exchange ring. In particular, all regular or semiperfect rings are exchange rings.

e. A right A-module X is called a *free cyclic* module if there exists an element $x \in X$ which is called a *free generator* for X such that $X = xA$ and the right annihilator $r(x)$ of the element x is equal to the zero. We note that X is a free cyclic module if and only if $X \cong A_A$.

A module X_A is said to be *free* if there exists a subset $\{x_i\}_{i \in I} \subseteq X$ which is called a *basis* of the module X such that $X = \oplus_{i \in I} x_i A$ and $r(x_i) = 0$ for all $i \in I$; the cardinality $\operatorname{card}(I)$

is called the *rank* of the free module X. We note that the rank of a free module is not necessarily unique.

A module X is said to be *finite-presented* if $X \cong F/N$, where F is a finitely generated free module and N is a finitely generated submodule in F.
A right module X over the ring A is said to be *cyclically presented* if $M \cong A_A/aA$ for some element $a \in A$.

f. A module X is said to be *projective with respect to* a module M if for any module epimorphism $h\colon M \to \overline{M}$ and each homomorphism $\overline{f}\colon X \to \overline{M}$, there exists a homomorphism $f\colon X \to M$ with $\overline{f} = hf$.

A module X is said to be *projective* if the following equivalent conditions hold.

1) X is projective with respect to any right A-module.
2) The module X is isomorphic to a direct summand of a free module.
3) The kernel of any epimorphism of an arbitrary A-module M onto the module X is a direct summand of the module M.

A module, which is projective with respect to itself, is called a *quasiprojective* or *self-projective* module.

It is clear that all projective modules are quasiprojective and cyclic group of any prime order is a quasiprojective nonprojective simple $\mathbb{Z}$-module.

A ring A is called a *right PP ring* or a *right Rickartian* ring if the following equivalent conditions hold.
1) For any element $x \in A$, the module xA is projective.
2) For any element $x \in A$, there exists an idempotent $e \in A$ with $r(x) = eA$.

A module is said to be *hereditary* if all its submodules are projective. As usual, a *hereditary* ring is a right and left hereditary ring.
A module is said to be *semihereditary* if all its finitely generated submodules are projective.

g. A ring A is called a *domain* (resp., *prime* ring) if the product of any two of its nonzero elements (resp., ideals) is not equal to the zero.
A proper right ideal P of the ring A is said to be *completely prime* if $ab \notin P$ for all $a, b \in A \setminus P$.
A proper ideal P of the ring A is said to be *completely prime* (resp., *prime*) if the factor ring A/P is a domain (resp., prime ring).
The intersection of all prime ideals of the ring A is a nil–ideal; it is called the *prime radical* of the ring A.
A ring A is said to be *semiprime* (resp., *reduced*) if A does not have nonzero nilpotent ideals (resp., nonzero nilpotent elements).

A ring without noncentral idempotents is called a *normal* or *Abelian* ring.

h. A module X is called an *essential extension* of its submodule Y if $Y \cap Z \neq 0$ for any nonzero (cyclic) submodule Z. In this case, Y is called an *essential submodule* in X. If X is a submodule of the module M and $X + Y \neq M$ for any proper submodule Y of M, then X is called a *small* submodule in M.

A module M is said to be *uniform* if any two of its nonzero submodules have the nonzero intersection, i.e., M is an essential extension of any of its nonzero submodules. A module X is said to be *finite-dimensional* if X does not contain infinite direct sums of nonzero (cyclic) submodules; X is finite-dimensional if and only if X is an essential extension of finite direct sum of nonzero uniform (cyclic) submodules.
A right finite-dimensional ring with the maximum condition on right annihilators is called a *right Goldie ring*.
Let n be a positive integer. One says that a module M has *uniform dimension n* or *Goldie dimensionn* if M is an essential extension of direct sum of n uniform of nonzero modules and M does not contain direct sum of $n + 1$ uniform of nonzero modules.

i. A nonzero module is said to be *simple* if it coincides with any of its nonzero submodules. Direct sums of simple modules are called *semisimple* modules. For a module M, we denote by $\operatorname{Soc} M$ the largest semisimple submodule of the module M; by definition, $\operatorname{Soc} M = 0$ if M does not contain semisimple submodules. $\operatorname{Soc} M$ is called the *socle* of the module M. Semisimple modules coincide with nonzero modules in which all submodules are direct summands. Right (left) semisimple rings coincide with rings that are isomorphic to finite direct products of matrix rings over division rings. All nonzero modules over semisimple Artinian rings are semisimple, injective, projective modules.

j. In this book, we also use other well-known notions and assertions of ring theory that are contained in many books; e.g., see [16, 63, 97, 131, 166, 194].

Contents

Part I: **ARITHMETICAL RINGS**

1 Saturated ideals and localizations

The main results of this section are Theorems 1A, 1B and 1C.

1A Theorem (Tuganbaev [172]). A right invariant ring A is arithmetical if and only if for any maximal ideal M, all $\{A \setminus M\}$-saturated ideals of the ring A form a chain with respect to inclusion.

1B Theorem (Jensen [98]). A commutative ring A is arithmetical if and only if for any maximal ideal M, the localization A_M is a uniserial ring.

1C Theorem (Tuganbaev [157]). A right invariant ring A is an arithmetical semiprime ring if and only if for any maximal ideal M, the right localization A_M exists and is a right uniserial domain.

Remark. The completion of the proof of Theorems 1A, 1B and 1(C) is given in 1.2.7 and 1.3.11.

1.1 Distributive modules

1.1.1 The first criterion of distributivity of the module. For a ring A and a right A-module X, the following conditions are equivalent.

1) X is a distributive module.

2) All submodules of all homomorphic images of the module X are distributive.

3) In X, all 2-generated submodules are distributive.

4) In X, every 2-generated submodule is contained in some distributive submodule of X.

5) $X_1 \cap (X_2 + X_3) = X_1 \cap X_2 + X_1 \cap X_3$ for any cyclic submodules X_1, X_2, X_3 in M.

6) A contains a unitary subring A' such that a natural A'-module X is distributive.

7) $(\sum_{i \in I} X_i) \cap (\sum_{j \in J} Y_j) = \sum_{i \in I, j \in J}(X_i \cap Y_j)$ for any two sets $\{X_i\}_{i \in I}$ and $\{Y_j\}_{j \in J}$ of submodules in M.

The equivalences from 1.1.1 are directly verified; they are used without specific references.

1.1.2 The second criterion of distributivity of a module ([127, 152]). For a ring A and a right A-module X, the following conditions are equivalent.

1) X is a distributive module.

2) For any two elements $x, y \in X$, there exists an element $a \in A$ such that $xaA + y(1 - a)A \subseteq xA \cap yA$.

https://doi.org/10.1515/9783110659825-001

3) For any two elements $x, y \in X$, there exist elements $a, b, c, d \in A$ such that $1 = a+b$, $xa = yc$ and $yb = xd$.

4) $A = (x^{\cdot}.\, yA) + (y^{\cdot}.\, xA)$ for any two elements $x, y \in X$.

Proof. The equivalences 2) $\Leftrightarrow$ 3) $\Leftrightarrow$ 4) are directly verified.

1) $\Rightarrow$ 2). Let $T = xA \cap yA$. Since $(x+y)A = (x+y)A \cap xA + (x+y)A \cap yA$, there exist $b, d \in A$ such that

$$(x+y)b \in xA\,, \quad (x+y)d \in yA\,, \quad x+y = (x+y)b + (x+y)d\,.$$

Therefore, $yb = (x+y)b - xb \in T$ and $xd = (x+y)d - yd \in T$. We set $a = 1 - b$ and $z = a - d = 1 - b - d$. Then

$$1 = a + b\,, \quad (x+y)z = (x+y) - (x+y)b - (x+y)d = 0\,,$$
$$xa = xd + xz = xd + (x+y)z - yz = xd - yz\,, \quad yz = -xz \in T\,, \quad xa \in T\,.$$

2) $\Rightarrow$ 1). Let X_1, X_2, X_3 be submodules in X and $x_3 = x_1 + x_2 \in (X_1 + X_2) \cap X_3$, where $x_1 \in X_1$ and $x_2 \in X_2$. By assumption, there exist two elements $a, b \in A$ such that $1 = a + b$, $x_1 a \in x_2 A$ and $x_2 b \in x_1 A$. Then

$$x_3 b = x_1 b + x_2 b \in xA \cap zA\,,$$
$$x_3 a = x_1 a + x_2 a \in yA \cap x_3 A\,, \quad x_3 = x_3 b + x_3 a \in X_1 \cap X_3 + X_2 \cap X_3\,,$$
$$(X_1 + X_2) \cap X_3 \subseteq X_1 \cap X_3 + X_2 \cap X_3 \subseteq (X_1 + X_2) \cap X_3\,.$$

The equivalence 1) $\Leftrightarrow$ 3) follows from the equivalence 1) $\Leftrightarrow$ 5). □

1.1.3 The third criterion of distributivity of a module. For a ring A and a right A-module M, the following conditions are equivalent.

1) M is a distributive module.

2) For any two elements $x, y \in M$, there exists a right ideal B in A such that $(x+y)A = xB + yB$.

Proof. 1) $\Rightarrow$ 2). We set $B \equiv (x^{\cdot}.\,(x+y)A)$. Then $(x+y)A \cap xA = xB$. If $b \in B$, then $yb = (x+y)b - xb \in (x+y)A$. Therefore,

$$B = (y^{\cdot}.\,(x+y)A)\,, \quad (x+y)A \cap yA = yB\,,$$
$$(x+y)A = (x+y)A \cap xA + (x+y)A \cap yA = xB + yB\,.$$

2) $\Rightarrow$ 1). Let $x, y \in M$. By assumption, $(x+y)A = xB + yB$, where B is a right ideal in A. It follows from the modular law that

$$xA \cap (x+y)A = xA \cap (xB + yB) = xB\,,$$
$$yA \cap (x+y)A = yA \cap (xB + yB) = yB\,,$$
$$(x+y)A = xA \cap (x+y)A + yA \cap (x+y)A$$

and M is distributive. □

1.1.4 The fourth criterion of distributivity of a module ([152]). For a ring A and a right A-module M, the following conditions are equivalent.

1) M is a distributive module.

2) $\operatorname{Hom}(X, Y) = 0$ for any subfactor $X \oplus Y$ of M.

3) M does not have subfactors that are direct sums of two isomorphic nonzero modules.

4) $\operatorname{Hom}(X/(X \cap Y), Y/(X \cap Y)) = 0$ for any two submodules X, Y of M.

5) $\operatorname{Hom}((X + Y)/Y, (X + Y)/X) = 0$ for any two submodules X, Y of M.

6) In every subfactor of the module M, any direct summand is fully invariant.

7) In the endomorphism ring of any subfactor of the module M, all idempotents are central.

Proof. The implication 2) $\Rightarrow$ 3) and the equivalence 2) $\Leftrightarrow$ 4) and 2) $\Leftrightarrow$ 6) $\Leftrightarrow$ 7) are directly verified.

The equivalence 4) $\Leftrightarrow$ 5) follows from natural isomorphisms $X/(X \cap Y) \cong (X + Y)/Y$ and $Y/(X \cap Y) \cong (X + Y)/X$.

3) $\Rightarrow$ 1). We assume that M is not distributive. There exist submodules X, Y, Z in M such that $X \subseteq Y + Z$ and $X/(X \cap Y + X \cap Z)$ is a nonzero submodule in $(Y + Z)/(X \cap Y + X \cap Z)$. Let $h\colon Y + Z \to (Y + Z)/(X \cap Y + X \cap Z)$ be the natural epimorphism. Then

$$0 \neq h(X) \subseteq h(Y + Z) = h(Y) \oplus h(Z)\,, \quad h(X) \cap h(Y) = 0\,, \quad h(X) \cap h(Z) = 0\,.$$

Let $f\colon h(Y) \oplus h(Z) \to h(Y)$ and let $g\colon h(Y) \oplus h(Z) \to h(Z)$ natural projections. Since $h(X) \cap h(Y) = 0$ and $h(X) \cap h(Z) = 0$, we have $f(h(X)) \cong h(X) \cong g(h(X)) \neq 0$. This is a contradiction.

1) $\Rightarrow$ 2). Let $f \in \operatorname{Hom}(X, Y)$, $x \in X$ and $y = f(x) \in Y$. According to 1.1.2, there exists an element $a \in A$ such that

$$xaA + y(1 - a)A \subseteq xA \cap yA \subseteq X \cap Y = 0\,,$$
$$xa = y(1 - a) = 0\,, \quad y = ya = f(x)a = f(xa) = f(0) = 0\,.$$

Therefore, $f \equiv 0$ and $\operatorname{Hom}(X, Y) = 0$. □

1.1.5 The fifth criterion of distributivity of a module ([43]). For a right module M over the ring A, the following conditions are equivalent.

1) M is a distributive module.

2) M does not have subfactors that are direct sums of two isomorphic simple modules.

3) Any 2-generated submodule X of the module M does not have the factor module $S \oplus T$ such that S and T are isomorphic simple modules.

Proof. The implication 1) $\Rightarrow$ 2) follows from 1.1.4.

The implication 2) $\Rightarrow$ 3) is obvious.

3) $\Rightarrow$ 2). We assume the contrary. There exists a submodule X_1 in M and a submodule Y_1 in X such that $X_1/Y_1 = S_1 \oplus T_1$, where S_1, T_1 are isomorphic simple modules. Let $h\colon X_1 \to X_1/Y_1$ be the natural epimorphism. There exist cyclic submodules S_2, T_2 of the module X_1 such that $S_1 = h(S_2)$ and $T_1 = h(T_2)$. We denote by X_2 the 2-generated submodule $S_2 + T_2$ of the module M. Then $h(X_2)$ is a direct sum of two isomorphic simple modules. This is a contradiction.

2) $\Rightarrow$ 1). We assume that the module M is not distributive. According to 1.1.4, M has a subfactor $X \oplus Y$ such that X, Y are nonzero modules and there exists an isomorphism $f\colon X \to Y$. The nonzero module X has a nonzero cyclic submodule N. The module N has a simple factor module N/T. Then Y has a simple subfactor $f(N)/f(T)$. Therefore, M has a subfactor that is isomorphic to the module $N/T \oplus N/T$. This is a contradiction. □

1.1.6 Distributive modules over invariant rings. Let A be a right invariant ring and let M be a distributive right A-module.

a. $A = (Y^{\cdot}.X) + (X^{\cdot}.Y)$ for any finitely generated submodules X, Y of the module M.

b. For any submodule Z' of an arbitrary finitely generated submodule Z of the module M, there exists an ideal A' of the ring A such that $ZA' = Z'$.

c. If M is a finitely generated module, then M is an invariant module.

Proof. **a.** Since $X+Y$ is a finitely generated module, there exist a positive integer n and elements $x_i \in X$, $y_i \in Y$, $1 \le i \le n$, such that $X+Y = (x_1+y_1)A + \cdots + (x_n+y_n)A$. Since the module M is distributive, $(X+Y)\bigcap Z = (X\bigcap Z) + (Y\bigcap Z)$ for any submodule Z in M.

Let $y \in Y$. For any $1 \le i \le n$, we have

$$(x_i+y)A = (x_i+y)A\bigcap(X+Y) = [(x_i+y)A\bigcap X] + [(x_i+y)A\bigcap Y]\,.$$

Therefore, there exist elements $a \in A$ and $z \in Y$ such that

$$(x_i+y)a \in X\,, \quad x_i + y = (x_i+y)a + z\,.$$

Therefore, $x_i(1-a) \in Y$ and $ya \in X$. Consequently,

$$A = (yA^{\cdot}.X) + (x_iA^{\cdot}.Y)\,, \quad 1 \le i \le n\,.$$

Therefore,

$$A = (yA^{\cdot}.X) + [(x_1A^{\cdot}.Y)\bigcap\cdots\bigcap(x_nA^{\cdot}.Y)] = (yA^{\cdot}.X) + (X^{\cdot}.Y)\,.$$

In particular,

$$A = (y_iA^{\cdot}.X) + (X^{\cdot}.Y) \quad (1 \le i \le n)\,.$$

Therefore,

$$A = [(y_1A\,^{\cdot}.\,X)\bigcap\cdots\bigcap(y_nA\,^{\cdot}.\,X)] + (X\,^{\cdot}.\,Y) = (Y\,^{\cdot}.\,X) + (X\,^{\cdot}.\,Y)\,.$$

b. Let Z be an n-generated module, $n \in \mathbb{N}$. We use induction on n. For $n = 1$, we can identify the cyclic A-module Z over the right invariant ring A with the right invariant factor ring $A/r(Z)$ of the ring A. In this case, the assertion is directly verified.

Now we assume that the assertion it is true for all k-generated submodules of the module M for $k < n$. We can assume that $Z = X + Y$, where X is a cyclic module and Y is a $(n-1)$-generated module. By the induction hypothesis, there exist ideals B and C of the ring A such that $X \cap Y = XB = YC$. Therefore, $X\bigcap Y = X(X\,^{\cdot}.\,Y) = Y(Y\,^{\cdot}.\,X)$ By **a**, $A = (Y\,^{\cdot}.\,X) + (X\,^{\cdot}.\,Y)$. Therefore,

$$X = X((Y\,^{\cdot}.\,X) + (X\,^{\cdot}.\,Y)) = X(Y\,^{\cdot}.\,X) + X(X\,^{\cdot}.\,Y) = X(Y\,^{\cdot}.\,X) + Y(Y\,^{\cdot}.\,X) = ZB\,,$$

where $B = (Y\,^{\cdot}.\,X)$. Similarly, we have $Y = ZC$, where $C = (X\,^{\cdot}.\,Y)$.

Let Z' be a submodule in $Z = X + Y$. We have to prove that there exists an ideal H of the ring A such that $Z' = (X + Y)H$. By assumption, $Z' = X \cap Z' + Y\bigcap Z'$. By the induction hypothesis, there exist ideals D, E of the ring A such that $Z'\bigcap X = XD$ and $Z'\bigcap Y = YE$. In addition, $X = ZB$ and $Y = ZC$. Therefore,

$$Z' = X\bigcap Z' + Y\bigcap Z' = XD + YE = ZBD + ZCE = Z(BD + CE)$$

and $BD + CE$ is the required ideal A' of the ring A.

c. By **b**, there exists an ideal B of the ring A with $X = MB$. Let f be an endomorphism of the module M. Then

$$f(X) = f(MB) \subseteq f(M)B \subseteq MB = X\,. \qquad \square$$

1.1.7 Distributive rings. Let A be a right distributive ring.

a. If X, Y are two right ideals of the ring A and $X \cap Y = 0$, then $(AX)(AY) = (AY)(AX) = 0$. Consequently, all idempotents of the ring A are central.

b. If A is a domain, then the ring A is right uniform.

c. If the ring A is right invariant and M is a right A-module, then $MB \cap MC = M(B \cap C)$ for any right ideals B, C in A.

Proof. **a.** We prove that $yX = 0$ for any element $y \in Y$. We define a homomorphism of right A-modules $f\colon X \to Y$ by the relation $f(x) = yx$ for any $x \in X$. By 1.1.4, $yx = f(x) = 0$; see condition 2) from 1.1.4. Therefore, $0 = YX = XAY$. Then $(AX)(AY) = 0$.

b. The assertion follows from **a**.

c. Let $x = \sum_{i=1}^{s} m_ib_i = \sum_{j=1}^{t} n_jc_j \in MB \cap MC$, where $m_i, n_j \in M$, $b_i \in B$ and $c_j \in C$. We set $B_1 = \sum_{i=1}^{s} b_iA \subseteq A$ and $C_1 \sum_{j=1}^{t} c_j \subseteq A$. Then B_1, C_1 are finitely generated right ideals of the right distributive, right invariant ring A. By 1.1.6(a), there exist two elements $a, b \in A$ such that $1 = a + b$ and $b_i a, c_j b \in B_1 \cap C_1$ for all i, j. Therefore,

$$x = xa + xb = \sum_{i=1}^{s} b_i a + \sum_{j=1}^{t} c_j b \in M(B_1 \cap C_1) \subseteq M(B \cap C)$$

and $MB \cap MC \subseteq M(B \cap C) \subseteq MB \cap MC$. □

1.2 Saturated submodules and saturations

1.2.1 Saturated submodules. Let A be a ring, X be a right A-module, X_1 be a submodule in X, S be a nonempty multiplicatively closed subset in A that contains $1 \in A$ and does not contain $0 \in A$, and let $\overline{X_1}_S$ be the subset in X consisting of all elements $x \in X$ such that $xs \in X_1$ for some element $s \in S$.
The set $\overline{X_1}_S$ is called the S*–saturation* of the set X_1. Since $1 \in S$, we have $X_1 \subseteq \overline{X_1}_S$.
If $X_1 = \overline{X_1}_S$, then X_1 is called an S*-saturated* submodule of the module X. It is clear that X_1 is an S-saturated submodule if and only if X_1 contains any element $x \in X$ such that $xs \in X_1$ for some element $s \in S$.
A subset B of the ring A is said to be *right permutable* if for any elements $a \in A$ and $b \in B$, there exist two elements $a' \in A$, $b' \in B$ with $ab' = ba'$.

a. If the set S is *right permutable*, then the S-saturation $\overline{X_1}_S$ is an S-saturated submodule in X, which contains X_1.

b. If the ring A is right invariant, then each of its subsets B are right permutable.

c. If S is a right permutable multiplicatively closed subset of the ring A, which contains $1 \in A$ and does not contain $0 \in A$, and Y is an arbitrary submodule of the module X, then its S-saturation $\overline{Y}$ is an S-saturated submodule in M and coincides with the intersection $\hat{Y}$ of all S-saturated subsets in X that contain Y.

Proof. **a.** Indeed, let $y, z \in \overline{X_1}_S$. Then $ys \in X_1$ and $zt \in X_1$ for some elements $s, t \in S$. Since S is right permutable, $ts' = st'$ for some $s' \in S$ and $t' \in A$. Since S is multiplicatively closed, $ts' \in S$. Then $(y+z)ts' = yst' + zts' \in X_1$. Therefore, $y+z \in \overline{X_1}_S$ and $\overline{X_1}_S$ is an additive subgroup in X_1.

Let $a \in A$. Since S is right permutable, $as' = sa'$ for some $s' \in S$ and $a' \in A$. Then $yas' = (ys)a' \in X_1$. Therefore, $\overline{X_1}_S$ is a submodule in X, which contains X_1.

It remains to prove that $x \in \overline{X_1}_S$ if $xu \in \overline{X_1}_S$ for some element $u \in S$. Since $xu \in \overline{X_1}_S$, we have that $x(uv) \in X_1$ for some element $u \in S$. Then $uv \in S$, since S is multiplicatively closed. □

b. Let $a \in A$ and $b \in B$. Since the right ideal bA is an ideal, $ba \in bA$ and $ab = ba'$ for some $a' \in A$.

c. Let $x, z \in \overline{Y}$ and $a \in A$. There exist $s, s' \in S$ such that $xs, zs' \in Y$. By assumption, $at = sb$ and $sb' = s't' \in S$ for some $t, t' \in S$ and $b, b' \in A$. Then

$$(xa)t = (xs)b \in Y\,, \quad (x+z)s't' = (xs)b' + (zs')t' \in Y\,, \quad xa, x+z \in Y^S\,.$$

Therefore, $\overline{Y}$ is a submodule in X. Since $1 \in S$, we have $Y \subseteq \overline{Y}$. Let $x \in X$ and $xs \in \overline{Y}$ for some $s \in S$. It follows from the definition of $\overline{Y}$ that $xst \in Y$ for some $t \in S$ and $st \in S$, since S is multiplicatively closed. Therefore, $x \in \overline{Y}$, whence $\overline{Y}$ is an S-saturated submodule in X. Now it follows from the definition of Y^S that $\overline{Y} = \hat{Y}$. □

1.2.2 Properties of distributive modules ([152]). Let A be a ring, X be a distributive right A-module, Y be a submodule in X, $f \in \operatorname{End} X$, and let $f^{-1}(Y) = \{x \in X \mid f(x) \in Y$.

a. If $X = Y + f^{-1}(Y)$, then $f(Y) \subseteq Y$.

b. If $f(X) \subseteq Y + f(Y)$, then $f(Y) \subseteq Y$.

c. If $Y \neq X$, then $Y + f(Y) \neq X$.

d. If $x_1A \oplus \cdots \oplus x_nA$ is a submodule in X, then $x_1A \oplus \cdots \oplus x_nA = (x_1 + \cdots + x_n)A$ is a cyclic module.

e. If the ring A is local, then X is a uniserial module.

Proof. **a.** We assume that $X = Y + f^{-1}(Y)$. The rule $g(x + f^{-1}(Y)) = f(x) + Y$ correctly defines a monomorphism $g\colon X/f^{-1}(Y) \to X/Y$. According to 1.1.4, $g \equiv 0$. Therefore, $f^{-1}(Y) = X$ and $f(X) \subseteq Y$.

b. We assume that $f(X) \subseteq Y + f(Y)$. Let $x \in X$. By assumption, $f(x) = y + f(z)$ for some $y, z \in Y$. Therefore, $x - z \in f^{-1}(Y)$, $x = z + (x - z) \in Y + f^{-1}(Y)$. Then $X = f^{-1}(Y) + Y$ and $f(X) \subseteq Y$ by **a.**

c. We assume that $X = Y + f(Y)$. Then $f(X) \subseteq Y + f(Y)$. By **a**, $f(X) \subseteq Y$. Then $X = Y + f(Y) \subseteq Y + f(X) = Y$. This is a contradiction.

d. Without loss of generality, we can assume that $n = 2$. According to 1.1.2, there exist two elements $a, b \in A$ such that $1 = a + b$ and $x_1aA + x_2bA \subseteq x_1A \cap x_2A = 0$. Therefore, $x_1a = x_2b = 0$. Then $x_1 = x_1(a + b) = (x_1 + x_2)b$ and $x_2 = x_2(a + b) = (x_1 + x_2)a$. Therefore, $x_1A + x_2A = (x_1 + x_2)A$ is a cyclic module.

e. Let $x, y \in X$. It is sufficient to prove that the submodules xA and yA are comparable with respect to inclusion. According to 1.1.2, there exist elements $a, b, c, d \in A$ such that $1 = a + b$ and $xaA + ybA \subseteq xA \cap yA$. Since the ring A is local and $1 = a + b$, at least one of the right ideals aA, bA coincides with A. Therefore, at least one of the inclusions $xA \subseteq yA$, $yA \subseteq xA$ is true. □

1.2.3 Quasi-invariant modules and rings. A module X is said to be *quasi-invariant* if all its maximal submodules are fully invariant in M.

a. A ring A is right quasi-invariant if all its maximal right ideals are ideals.

b. If a ring A is right quasi-invariant and M is its maximal right ideal, then M is a maximal ideal, the factor ring A/M is a division ring and $\{A \setminus M\}$ is a nonempty multiplicatively closed subset in A which contains the identity element of the ring A and does not contain its zero.

c. Let a ring A be right quasi-invariant and let M be its maximal right ideal such that the set $\{A \setminus M\}$ is right permutable. Then M is a maximal ideal, the factor ring A/M is a division ring and $\{A \setminus M\}$ is a right permutable, nonempty, multiplicatively closed subset in A, which contains the identity element of the ring A and does not contain its zero and for any right A-module X and an arbitrary submodule X_1 in X, the $\{A \setminus M\}$-saturation $\overline{X_1}_{\{A\setminus M\}}$ is an $\{A \setminus M\}$-saturated submodule in X, which contains X_1.

d. Every distributive module X is quasi-invariant.

e. Every right distributive ring A is right quasi-invariant and for any its maximal (right) ideal M, it is true that the set $\{A \setminus M\}$ is right permutable.

Proof. **a, b.** The assertions are directly verified.

c. The assertion follows from **b** and 1.2.1(a).

d. Let Y be a maximal submodule in X. We assume that $f(Y) \not\subseteq Y$. Then $X = Y + f(Y)$. This contradicts to 1.2.2(c).

e. By **d**, the ring A is right quasi-invariant. Let M be a maximal (right) ideal of the ring A and $\{A \setminus M\}$. It remains to prove that for any elements $x \in A$ and $y \in \{A \setminus M\}$, there exist elements $x' \in A$ and $y' \in \{A \setminus M\}$ with $xy' = yx'$. By 1.1.2, there exist elements $a, c, d \in A$ such that $xa = yc$ and $y(1 - a) = xd$. If $a \in \{A \setminus M\}$, then we can set $y' = a$ and $x' = c$. Therefore, we can assume that $a \in M$. Then $1 - a \in \{A \setminus M\}$ and $xd = y(1 - a) \in \{A \setminus M\}$. Since M is an ideal, $d \in \{A \setminus M\}$. In this case, we can set $y' = d$ and $x' = 1 - a$. □

1.2.4 A criterion of distributivity of the module over a quasi-invariant ring with permutable complements of maximal ideals. Let A be a right quasi-invariant ring, X be a right A-module, and for any maximal (right) ideal M of the ring A, the (multiplicatively closed) set $\{A \setminus M\}$ is right permutable. Then the following conditions are equivalent.

1) X is a distributive module.

2) For any maximal (right) ideal M of the ring A, all $\{A \setminus M\}$-saturated submodules of the module X form a chain with respect to inclusion.

Proof. 1) ⇒ 2). Let M be a maximal ideal of the ring A and let Y, Z be two $\{A \setminus M\}$-saturated submodules of the module X. We assume that Y is not contained in Z. Then there exists an element $y \in Y \setminus Z$. We have to prove that $z \in Y$, where z is an arbitrary element of Z. According to 1.1.2, there exist two elements $a, b \in A$ such that $1 = a + b$ and $yaA + zbA \subseteq Y \cap Z$.

Let $a \in \{A \setminus M\}$. Since the element ya is contained in the $\{A \setminus M\}$-saturated submodule Z, we have $y \in Z$; this contradicts to the choice of $y \in Y \setminus Z$.

Therefore, $a \in M$. Then $b = 1 - a \in \{A \setminus M\}$. Since the element zb is contained in the $\{A \setminus M\}$-saturated submodule Y, we have $z \in Y$.

2) $\Rightarrow$ 1). Let $x_1, x_2 \in X$. According to 1.1.2, it is sufficient to prove that $A = (x_1 \,{}^{\cdot}\!.\, x_2A) + (x_2 \,{}^{\cdot}\!.\, x_1A)$.

We assume that $A \neq (x_1 \,{}^{\cdot}\!.\, x_2A) + (x_2 \,{}^{\cdot}\!.\, x_1A)$. Then the proper right ideal $(x_1 \,{}^{\cdot}\!.\, x_2A) + (x_2 \,{}^{\cdot}\!.\, x_1A)$ is contained in some maximal (right) ideal M. We denote by $\overline{x_1A}$ and $\overline{x_2A}$ the $\{A \setminus M\}$-saturations in X of the submodules x_1A and x_2A, respectively. According to 1.2.3(c), $\overline{x_1A}$ and $\overline{x_2A}$ are $\{A \setminus M\}$-saturated submodules in X. From 2), $\overline{x_1A}$ and $\overline{x_2A}$ are comparable with respect to inclusion. For example, let $\overline{x_1A} \subseteq \overline{x_2A}$. Since $x_1A \subseteq \overline{x_1A} \subseteq \overline{x_2A}$, we have that $x_1s \in x_2A$ for some $s \in \{A \setminus M\}$. Therefore, $s \in (x_1 \,{}^{\cdot}\!.\, x_2A) \subseteq M \cap \{A \setminus M\} = \varnothing$. This is a contradiction. □

1.2.5 A criterion of distributivity of a ring. A ring A is right distributive if and only if for any its maximal[1] (right) ideal M, all $\{A \setminus M\}$-saturated right ideals of the ring A form a chain with respect to inclusion.

The assertion 1.2.5 follows from 1.2.3(e) and 1.2.4.

1.2.6 Remark. Let A be an arbitrary commutative ring, M be its maximal ideal, A_M be the localization of the ring A by M, X be an A-module, and let X_M be the localization of X by M. It is well known that any submodule X'_M of the A_M-module X_M corresponds to the unique $\{A \setminus M\}$-saturated submodule X' of the A-module X such that its localization by M coincides with X'_M; proper inclusions are preserved under this correspondence.

1.2.7 The completion of the proof of Theorems 1A and 1B. Theorem 1A follows from 1.2.5 and the property that any right invariant arithmetical ring is right distributive. Theorem 1B follows from Theorem 1A and Remark 1.2.6.

1.3 Localizable rings

1.3.1 Reduced rings. Let A be a reduced ring (i.e., a ring without nonzero nilpotent elements).

a. If $a_1, \dots, a_n \in A$ and $a_1 \cdot \dots \cdot a_n = 0$, then $Aa_{s(1)}A \cdot \dots \cdot Aa_{s(n)}A = 0$ for any permutation s of the set $\{1, \dots, n\}$.

b. If $x, y \in A$, then $xy = 0 \Leftrightarrow yx = 0 \Leftrightarrow (AxA)(AyA) = (AyA)(AxA) = 0 \Leftrightarrow (AxA) \cap (AyA) = 0$.

c. $xy = 0 \Leftrightarrow x^nAy = xAy^n = 0$ for all $n \in \mathbb{N}$. In particular, $r(a) = r(a^n)$ for all $a \in A$ and $n \in \mathbb{N}$.

1 We recall that the ring A is right quasi-invariant by 1.2.3(e).

d. For any subset $B \subseteq A$, the annihilators $r(B)$ and $\ell(B)$ coincide; in addition, they are ideals and

$$r(B) = \ell(B) = r(ABA) = \ell(ABA) = \{a \in A \mid ABA \cap AaA = 0\}\,.$$

e. A is a ring without noncentral idempotents and every right or left invertible element of the ring A is invertible.

f. For any subset B in A, it is true that the factor ring $A/r(B)$ is reduced and $h(ABA)$ is an essential right ideal of the ring $A/r(B)$, where $h\colon A \to A/r(B)$ is the natural epimorphism.

g. If A is a right uniform ring, then A is a domain.

h. In the ring A, every principal right ideal B is a quasiprojective right A-module and a free cyclic right module over the reduced ring $A/r(B)$.

Proof. **a, b.** The element $b = a_{s(1)} \cdot \cdots \cdot a_{s(n)}$ can be obtained from the product of $a_1 \cdot \cdots \cdot a_n$ by using finitely many permutations of two adjacent factors. If $x, y \in A$ and $xy = 0$, then

$$(yAx)^2 = 0\,, \quad yAx = 0\,, \quad (xAy)^2 = 0\,, \quad xAy = 0\,.$$

Therefore, $Aa_{s(1)}A \cdot \cdots \cdot Aa_{s(n)}A = 0$.

c. Since A is reduced, the relations $(AxA) \cap (AyA) = 0$ and $(AxA)(AyA) = 0$ are equivalent.

Let $a \in A$ and $b \in r(a)$. Then $a^n b^n = 0$. By the above, $(ab)^n = 0$. Therefore, $ab = 0$ and $r(a^n) \subseteq r(a) \subseteq r(a^n)$. Therefore, $xy = 0 \Leftrightarrow x^n Ay = xAy^n = 0$ for all $n \in \mathbb{N}$.

d. The assertion follows from **b**.

e. If $e = e^2 \in A$, then $e(1-e) = 0$. By **b**, $(AeA) \cap (A(1-e)A) = 0$. Therefore, e is a central idempotent.

Let $a, b \in A$ and $ab = 1$. Then ba is an idempotent. It is proven above that ba is a central idempotent. Therefore, $b = b(ab) = b(ba)$. Then $ba = a(bba) = ab = 1$.

f. By **d**, $r(B)$ is an ideal, Therefore, the factor ring $A/r(B)$ exists. Let $y \in A$ and $y^n \in r(B)$ for some positive integer n. Then $By^n = 0$. By **c**, $By = 0$ and $y \in r(B)$. Therefore, $A/r(B)$ is a reduced ring. We assume that $h(C) \cap h(AXA) = h(0)$, where C is some right ideal of the ring A, which contains $r(B)$. Since $h(ABA)$ is an ideal of the ring $h(A)$, we have

$$h(C)h(ABA) \subseteq h(C) \cap h(ABA) = h(0)\,.$$

Therefore, $CB \subseteq r(B)$ and $C \subseteq \ell(B)$. By **b**, $\ell(B) = r(B)$. Therefore, $C \subseteq r(B)$ and $h(C) = h(0)$. Therefore, $h(ABA)$ is an essential right ideal of the ring $A/r(B)$.

g. Let $0 \neq x, y \in A$. Since the ring A is right uniform, $xA \cap yA \neq 0$. By **b**, $xy \neq 0$.

h. Let $B = bA$, $b \in B$. We denote by X the right annihilator of the element b in the ring A. By **d**, $X = r(B)$ is an ideal of the ring A. Let $h\colon A \to A/X$ be the natural ring epimorphism and let y be an element of the ring A such that $h(b)h(y) = h(0)$. Then $by \in X$ and $b^2y = 0$. Therefore, $(byb)^2 = 0$. Then $byb = 0$, whence $(by)^2 = 0$. Therefore, $by = 0$, $y \in X$, $h(y) = 0$, the element $h(b)$ has the zero right annihilator in the ring A/X, and the principal right ideal B is a free cyclic right A/X-module. In particular, B is a projective right A/X-module. Then it is directly verified that B is a quasiprojective right A/X-module. □

1.3.2 Right Ore sets. Let A be a ring and let S be a *right Ore set* in A, i.e., S is a right permutable multiplicatively closed subset in A, which contains 1 and does not contain 0 of the ring A, and all elements $s \in S$ are nonzero-divisors in A. In this case, it is well known that A is a unitary subring in the unique subring AS^{-1} such that all elements of S are invertible in AS^{-1} and $AS^{-1} = \{as^{-1} \mid a \in A,\ s \in S\}$. The ring AS^{-1} is called the *right ring of fractions* of the ring A with respect to right Ore set S.

If the set of all nonzero-divisors of the ring A is a right Ore set, then A is called a *right Ore ring* and the corresponding right ring of fractions is called the *right classical ring of fractions* of the ring A; it is denoted by $Q_{cl}(A)$. In this case, the ring A is called a *right order* in the ring $Q_{cl}(A)$.

For any subset B in A, we denote by BS^{-1} the subset $\{bs^{-1} \mid b \in B,\ s \in S\}$ of the ring AS^{-1}.

a. If $q_1, \ldots, q_n s_n^{-1} \in AS^{-1}$, then there exist elements $s \in A$ and $a_1, \ldots, a_n \in\in A$ such that $q_i = a_i s^{-1}$, $i = 1, \ldots, n$.
Consequently, the mapping $\varphi\colon \sum_{i=1}^{n} Aq_i$, $f(x) = xs$ defines an isomorphism from the left A-module $\sum_{i=1}^{n} Aq_i$ onto the n-generated left ideal $(\sum_{i=1}^{n} Aq_i)s$ of the ring A.

b. If A is a reduced ring, then AS^{-1} is a reduced ring.

c. The ring A is right uniform if and only if AS^{-1} is a uniform right A-module.

d. The ring A is right uniform if and only if the ring AS^{-1} is right uniform. Under these conditions, if A is a reduced ring, then A, AS^{-1} are of the domain.

e. If the ring A is right distributive, then the ring AS^{-1} is right distributive.

f. If A is a right distributive ring and AS^{-1} is a local ring, then AS^{-1} is a right uniserial ring, A is a right uniform ring and the right A-module AS^{-1} is uniform.

g. If A is a right distributive reduced ring and AS^{-1} is a local ring, then AS^{-1} is a right uniserial domain, A is a right uniform domain, and the right A-module AS^{-1} is uniform.

h. If N is a right ideal of the ring A and $q \in NS^{-1}$, then NS^{-1} is a right ideal of the ring AS^{-1} and $qs \in N$ for some $s \in S$.

i. If X is a right ideal of the ring AS^{-1}, then $X \cap A$ is a right ideal of the ring A and $X = (X \cap A)S^{-1}$.

j. If B, C are two right ideals of the ring A, then $(B+C)S^{-1} = BS^{-1}+CS^{-1}$ and $(B\cap C)S^{-1} = BS^{-1} \cap CS^{-1}$.

k. If $S = \{A \setminus M\}$, where M is a proper right ideal of the ring A, then the ring A_S is local and $J(A_S) = M_S$.

Under these conditions, if the ring A is right distributive, then A_S is a right uniserial ring, A is a right uniform ring, and the right A-module AS^{-1} is uniform.

Proof. **a.** It is sufficient to consider the case $n = 2$. There exist elements $d_i \in A$ and $s_i \in S$ such that $q_i = d_i s_i^{-1}$, $i = 1, 2$. Since S is right permutable, there exist elements $t_1 \in S$ and $t_2 \in A$ such that $s_1 t_1 = s_2 t_2 \in S$. We set $s = s_1 t_1 = s_2 t_2 \in S$ We set $a_i = d_i t_i \in A$, $i = 1, 2$. Then

$$s_i^{-1} = t_i s^{-1}, \quad a_i s^{-1} = d_i t_i s^{-1} = d_i s_i^{-1}(s_i t_i s^{-1}) = q_i, \quad i = 1, 2.$$

b. Let $as^{-1} \in AS^{-1}$ and $(as^{-1})(as^{-1}) = 0$, where $a \in A$ and $s \in S$. It is sufficient to prove that $as^{-1} = 0$. Since $s^{-1}a = bt^{-1}$ for some $b \in A$ and $t \in S$, it follows from the relation $(as^{-1})(as^{-1}) = 0$ that $ab = 0$. By 1.3.1(b), $(AxA) \cap (AyA) = 0$. On the other hand, it follows from the relation $s^{-1}a = bt^{-1}$ that $at = sb \in (AxA) \cap (AyA) = 0$. Therefore, $s^{-1}a = bt^{-1} = 0$, Then $a = 0$ and $as^{-1} = 0$.

c. If AS^{-1} is a uniform right A-module, then its submodule A_A is also uniform.

Conversely, let the ring A be right uniform and let $\bar{x}, \bar{y}$ be nonzero elements of the ring AS^{-1}. According to **a**, there exist nonzero elements $x, y \in A$ and $s \in S$ such that $\bar{x} = xs^{-1}$ and $\bar{y} = ys^{-1}$. Since x, y are two nonzero elements of the right uniform ring A, we have that $xa = yb \neq 0$ for some elements $a, b \in A$. Therefore, $\bar{x}sa = xa = yb = \bar{y}sb \neq 0$ and AS^{-1} is a uniform right A-module.

d. If the ring A is right uniform, it follows from **c** that the right A-module AS^{-1} is uniform. Therefore, the ring AS^{-1} is right uniform.

Conversely, let the ring AS^{-1} be right uniform and $0 \neq a, b \in A$. There exist elements $\bar{x}, \bar{y} \in AS^{-1}$ such that $a\bar{x} = b\bar{y} \neq 0$. According to **a**, there exist nonzero elements $x, y \in A$ and $s \in S$ such that $\bar{x} = xs^{-1}$ and $\bar{y} = ys^{-1}$. Then $ays^{-1} = bys^{-1} \neq 0$. Therefore, $ay = by \neq 0$ and the ring A is right uniform.

The remaining assertion follows from 1.3.1(f).

e. Let $\bar{x}, \bar{y} \in AS^{-1}$. By 1.1.2, it is sufficient to prove that there are elements $\bar{a}, \bar{b} \in AS^{-1}$ such that

$$1 = \bar{a} + \bar{b}, \quad \bar{x}\bar{a}AS^{-1} + \bar{y}\bar{b}AS^{-1} \subseteq \bar{x}AS^{-1} \cap \bar{y}AS^{-1}.$$

According to **a**, there exist elements $x, y \in A$ and $s \in S$ such that $\bar{x} = xs^{-1}$ and $\bar{y} = ys^{-1}$. Since the ring A is right distributive, it follows from 1.1.2 that there exist elements $a, b \in A$ such that $1 = a + b$, $xaA + ybA \subseteq xA \cap yA$. We set $\bar{a} = sas^{-1}$, $\bar{b} = sbs^{-1}$. Then

$$1 = s(a + b)s^{-1} = sas^{-1} + sbs^{-1} = \bar{a} + \bar{b},$$

$$\begin{aligned}\bar{x}\bar{a}AS^{-1} + \bar{y}\bar{b}AS^{-1} &= (xs^{-1})(sas^{-1})AS^{-1} + (ys^{-1})(sbs^{-1})AS^{-1} \\ &= xaAS^{-1} + ybAS^{-1} \subseteq (xA \cap yA)AS^{-1} \subseteq \bar{x}AS^{-1} \cap \bar{y}AS^{-1}.\end{aligned}$$

f. According to **e**, the ring AS^{-1} is right distributive. From 1.2.2(e), the right distributive local ring AS^{-1} is a right uniserial ring. From **c** and **d**, the ring A is right uniform and the right A-module AS^{-1} is uniform.

g. By **b**, AS^{-1} is a reduced ring. From **f**, it is sufficient to prove that the right uniserial reduced ring AS^{-1} is a domain; this follows from 1.3.1(f).

h. If $q \in BS^{-1}$, then it is clear that $qs \in B$ for some $s \in S$.

We prove that NS^{-1} is a right ideal of the ring A_S. Let $\bar{x}, \bar{y} \in NS^{-1}$ and $\bar{x} = xs_1^{-1}$, $\bar{y} = ys_2^{-1}$, where $x, y \in N$ and $s_1, s_2 \in S$. When assertion **a** is applied to the elements s_1^{-1}, s_2^{-1}, there exist elements $a, b \in A$ and $s \in S$ such that $s_1^{-1} = as^{-1}$ and $s_a^{-1} = bs^{-1}$. Then $\bar{x} = (xa)s^{-1}$ and $\bar{y} = (yb)s^{-1}$, where $xa, yb \in N$. Therefore, $\bar{x} + \bar{y} \in NS^{-1}$.
It remains to prove that $\bar{x}z = x(s_1^{-1}z) \in M_S$ for any $z \in AS^{-1}$. This follows from the inclusions $x \in M$ and $s_1^{-1}z \in AS^{-1}$.

i, j. The assertions are verified with the use of **a**.

k. Let $as^{-1} \in J(AS^{-1})$, where $a \in A$ and $s \in S$. The inclusion $a \in S$ is impossible; otherwise, the invertible element as^{-1} of the ring A_S is contained in $J(A_S)$. Therefore, $a \in A \setminus \{A \setminus M\}$, $as^{-1} \in MS^{-1}$ and $J(AS^{-1}) \subseteq MS^{-1}$.

Now let $mt^{-1} \in M_S$, where $m \in M$ and $t \in S$. Then $t - m \in S$, since otherwise $t - m \in M$ and $t = (t - m) + m \in M \cap S = \varnothing$. Then the element $1 - mt^{-1} = (t - m)t^{-1}$ is invertible in A and MS^{-1} is a right ideal of the ring AS^{-1} by **h.** Therefore, $MS^{-1} \subseteq J(A_S) \subseteq MS^{-1}$ and every element of $J(AS^{-1}) = MS^{-1}$ is invertible in AS^{-1}. This implies that the ring A_S is local and $J(A_S) = M_S$.

We assume that the ring A is right distributive. By **f**, AS^{-1} is a right uniserial ring, A is a right uniform ring, and the right A-module AS^{-1} is uniform. □

1.3.3 Denominator sets and reversive sets. Let A be a ring and S a nonempty subset in A.

a. Reversive and weakly reversive sets.
The set S is said to be *right reversive* if for any $a \in A$ and $s \in S$ with $sa = 0$, there exists an element $s' \in S$ with $as' = 0$.
The set S is said to be *is weakly right reversive* if for any $a \in A$ and $s \in S$ with $sa = a^2 = 0$, there exists an element $s' \in S$ with $as' = 0$.
(*Weakly*) *left reversive* sets are similarly defined.

b. Rings of fractions.
The set S is called a *right denominator set* in the ring A if there exist a nonzero ring AS^{-1} and a ring homomorphism $f_S \equiv f\colon A \to AS^{-1}$ such that all elements of $f(S)$ are invertible in AS^{-1}, $AS^{-1} = \{f(a)f(s)^{-1} \mid a \in A,\ s \in S\}$, and $\operatorname{Ker} f = \{a \in A \mid as = 0$ for some $s \in S\}$. In this case, AS^{-1} is called the *right ring of fractions* of A with respect to S and f_S is called the *canonical ring homomorphism* for AS^{-1}.
The definition of the ring AS^{-1} is consistent with the definition from 1.3.2 of the right ring of fractions with respect to a right Ore set.

c. The definitions of a *left denominator set* S, the *left ring of fractions* $S^{-1}A$ and the *canonical ring homomorphism* ${}_Sf\colon A \to {}_SA$ is similar to the definitions from **b.**
The set S is called a *two-sided denominator set* if there exist a ring $S^{-1}AS^{-1}$ and a ring homomorphism $f\colon A \to S^{-1}AS^{-1}$ such that all elements of $f(S)$ are invertible in the ring $S^{-1}AS^{-1}$, $S^{-1}AS^{-1} = \{f(a)f(s)^{-1} \mid a \in A,\ s \in S\} = \{f(t)^{-1}f(a) \mid a \in A,\ t \in S\}$, and $\operatorname{Ker} f = \{a \in A \mid as = ta = 0$ for some $s, t \in S\}$. In this case, $S^{-1}AS^{-1}$ is called the *two-sided ring of fractions* of the ring A with respect to S and the homomorphism f is called the *canonical homomorphism* for $S^{-1}AS^{-1}$.

d. In any ring, it is directly verified that every central multiplicatively closed subset, which contains $1 \in A$ and does not contain $0 \in A$, is a reversive, permutable, two-sided denominator set.

1.3.4 An existence criterion of AS^{-1}. Let A be a ring and let S be a multiplicatively closed subset in A, which contains $1 \in A$ and does not contain $0 \in A$. We set $K(S) = \{a \in A \mid as = 0$ for some $s \in S\}$. The following conditions are equivalent.

1) The right ring of fractions AS^{-1} exists (i.e., S is a right denominator set).

2) $K(S)$ is an ideal in A and the set $h(S)$ is a right Ore set in the ring $h(A)$, where $h\colon A \to A/K(S)$ is the natural epimorphism.

3) S is a right permutable, right reversive set.

4) S is a right permutable, right weakly reversive set.

Proof. The equivalence of 1) and 2) follows from 1.3.2. The implications 3) ⇒ 4) and 3) ⇒ 2) are directly verified.

3) ⇒ 4). Since $K(S)$ is an ideal in A, we have that S is right reversive. Let $a \in A$ and $s \in S$. Since $h(S)$ is a right Ore set in $h(A)$, we have that $h(at_1) = h(sb_1)$ for some $b_1 \in A$ and $t_1 \in S$. Therefore, $at_1 - sb_1 \in K(S)$ and $(at_1 - sb_1)t_2 = 0$ for some $t_2 \in S$. We set $s' = t_1t_2 \in S$ and $b = b_1t_2$. Then $as' = sb$. Therefore, S is right permutable.

4) ⇒ 3). Let $a \in A$, $s \in S$ and $sa = 0$. We set $b = as$. Then $b^2 = sb = 0$. By assumption, $x = 0$ for some $x \in S$. We set $s' = sx \in S$. Then $as' = 0$ and S is right reversive. ☐

1.3.5 Properties of rings of fractions. Let A be a ring, S be a right denominator set in A, AS^{-1} be the right ring of fractions, and let $f\colon A \to AS^{-1}$, $h\colon A \to A/\operatorname{Ker} f$ be the canonical ring homomorphisms. For any subset B of the ring A, we denote by B_S the set $\{f(b)f(s)^{-1} \mid b \in B,\ s \in S\}$.

a. The ring AS^{-1} coincides with the right ring of fractions $h(A)h(S)^{-1}$ of the ring $h(A)$ with respect to its right Ore set $h(S)$ and $f = \bar{f}h$, where $\bar{f}\colon h(A) \to h(A)h(S)^{-1}$ is an embedding of the ring $h(S)$ in its right ring of fractions $h(A)h(S)^{-1}$ with respect to the right Ore set $h(S)$.

b. If N is a right ideal of the ring A and $q \in N_S$, then N_S is a right ideal of the ring A_S and $qs \in f(N)$ for some $s \in S$. Under these conditions, if $q = h(a) \in h(A)$, then $at \in N$ for some $t \in S$.

c. If $q_1, \ldots, q_n s_n^{-1} \in AS^{-1}$, then there exist elements $s \in A$ and $a_1, \ldots, a_n \in\in A$ such that $q_i = a_i s^{-1}$, $i = 1, \ldots, n$.

d. If X is a right ideal of the ring AS^{-1}, then $X \cap h(A)$ is a right ideal of the ring $h(A)$ whose complete pre-image X' in A is a right ideal of the ring A and $X = X'_S$.

e. If B, C and D are three right ideals of the ring A, then $(B + C)_S = B_S + C_S$, $(B \cap C)_S = B_S \cap C_S$, $((B + C) \cap D)_S = (B_S + C_S) \cap D_S$, $(B \cap D + C \cap D)_S = B_S \cap D_S + C_S \cap D_S$.

f. Let $S = \{A \setminus M\}$, where M is a proper right ideal of the ring A and $\operatorname{Ker} f \subseteq M$. Then the ring A_S is local and $J(A_S) = M_S$.

Proof. **a.** The assertion follows from 1.3.4.

b. Since N is a right ideal of the ring A, we have that $h(N)$ is a right ideal of the ring $h(A)$ and $AS^{-1} = h(A)h(S)^{-1}$ by **a.** From 1.3.2(h), N_S is a right ideal of the ring A_S and $qs \in f(N)$ for some $s \in S$. We assume that $q = h(a) \in h(A) \cap N_S$, where $a \in A$. Then $as \in N + \operatorname{Ker} f$ for some $s \in S$. Then $ass' \in N$ for some $s' \in S$. Therefore, $at \in S$, where $t = ss' \in S$.

c, f. The assertions are verified with the use of **a** and 1.3.2(a),(k).

d, e. The assertions are verified with the use of **a** and 1.3.2(h),(i),(j). □

1.3.6 Rings of fractions of reduced rings. Let A be a ring and let S be a right permutable, multiplicatively closed subset in A, which contains $1 \in A$ and does not contain $0 \in A$. We set $K(S) = \{a \in A \mid as = 0 \text{ for some } s \in S\}$.

a. S is a right denominator set in A and the kernel canonical ring homomorphism $f\colon A \to AS^{-1}$ coincides with $K(S)$. If $h\colon A \to A/\operatorname{Ker} f$ is the canonical ring epimorphism, then $h(A)$ is a reduced ring and $h(A)h(S)^{-1} = AS^{-1}$ is a right ring of fractions rings $h(A)$ with respect to right Ore set $h(S)$.

b. Let $S = \{A \setminus M\}$, where M is a proper right ideal of the ring A which contains $\operatorname{Ker} f$. If A is a right distributive ring, then AS^{-1} is a right uniserial domain, $J(A_S) = M_S$, $A/\operatorname{Ker} f$ is a right uniform domain and the right A-module AS^{-1} is uniform.

Proof. **a.** Since A is a reduced ring, the set S is right weakly reversive. By 1.3.4, S is a right denominator set and $K(S) = \operatorname{Ker} f$ is an ideal of the ring A. By 1.3.4, $h(A)h(S)^{-1} = AS^{-1}$ is the right ring of fractions of the ring $h(A)$ with respect to the right Ore set $h(S)$. If $a \in A$ and $a^2 \in K(S)$, then $a^2 s = 0$ for some $s \in S$. By 1.3.1(c), $as = 0$, $a \in K(S)$ and $h(A)$ is a reduced ring.

b. The assertion follows from **a** and 1.3.2(g),(k). □

1.3.7 Prelocalizable and localizable rings. Localizations. A ring A is said to be *right prelocalizable* if every its maximal right ideal M is an ideal and the set $\{A \setminus M\}$ is right permutable. Then $\{A \setminus M\}$ is a nonempty set which contains $1 \in A$ and does not contain $0 \in A$. Since the factor ring A/M is a division ring, the set $\{A \setminus M\}$ is multiplicatively closed.

A ring A is said to be *right localizable* if A is right prelocalizable and for any its maximal right ideal M, the set $\{A \setminus M\}$ is right weakly reversive. It follows from 1.3.4 that for any maximal right ideal M of the right localizable ring A, there exists the right ring of fractions $A_{\{A\setminus M\}}$, which is called the **right localization** of the ring A by M; it is denoted by A_M, similar to the classical commutative case.

a. It is directly verified that every commutative ring is localizable and every right invariant ring is right prelocalizable.

b. Every right distributive ring is right prelocalizable by 1.2.3(e).

c. The assertion V from the introduction contains an example of a prelocalizable, invariant, distributive ring which is not right or left localizable.

d. If the ring A is right prelocalizable and for any its maximal ideal M, the set $\{A \setminus M\}$ is right weakly reversive, then the ring A is right localizable by 1.3.4.

e. Let the ring A be right invariant. If for any its maximal ideal M, the set $\{A \setminus M\}$ is right weakly reversive (e.g., this is the case if A is a reduced ring), then the ring A is right localizable by **(a)** and **(d)**.

f. Let the ring A is right distributive and for any its maximal ideal M, the set $\{A \setminus M\}$ is right weakly reversive. Then the ring A is right localizable by **(b)** and **(d)**. By 1.2.2(d), A_M is a right uniserial ring for any $M \in \max A$ and $J(A_M) = M_M$.

1.3.8 Properties of localizable rings. Let A be a right localizable ring and let $\max A$ be the set of all its maximal (right) ideals. For any $M \in \max A$ and each subset B in A, we denote by f_M and B_M the canonical ring homomorphism $A \to A_M$ and the set $\{f_M(b)f_M(s)^{-1} \mid b \in B,\ s \in A \setminus M\}$, respectively.

a. For any $M \in \max A$, the right localization A_M is a local ring and $J(A_M) = M_M$.

b. Let $a \in A$ and N a right ideal of the ring A. The element a is contained in N if and only if the element a_M is contained in the right ideal N_M of A_M for any $M \in \max A$. In particular, $a = 0$ if and only if $a_M = 0_M$ for any $M \in \max A$.

c. If N, N' are two right ideals of the ring A, then the relation $N = N'$ is equivalent to the property that $N_M = N'_M$ for any $M \in \max A$.

d. The ring A is right distributive if and only if for every $M \in \max A$, the ring A_M is right distributive, if and only if for every $M \in \max A$, the ring A_M is right uniserial (under the last condition, the relation $J(A_M) = M_M$ is true).

e. A is a reduced ring if and only if A_M is a reduced ring for every $M \in \max A$.

Proof. **a.** The assertion follows from 1.3.5(f).

b. If $a \in N$, then it is clear that $a_M \in N_M$ for any $M \in \max A$.

We assume that $a \notin N$. Then $(a \,^{\cdot}_{\cdot}\, N)$ is a proper right ideal in A, contained in some $M \in \max A$. By assumption, $a_M \in B_M$. By 1.3.5(b), $as \in N$ for some $s \in S$. Therefore, $(a \,^{\cdot}_{\cdot}\, N)$ is not contained in M. This is a contradiction.

c. The assertion follows from **b.**

d. The first equivalence of follows from 1.3.5(d),(e). The second equivalence follows from the first equivalence and 1.3.7(f).

e. The assertion is verified by the use of the second assertion from **b.** □

1.3.9. For a ring A, the following conditions are equivalent.
1) A is a right distributive reduced ring.
2) The ring A is right localizable and A_M is a right uniserial domain for every $M \in \max A_A$.
Under the conditions 1) and 2), $J(A_M) = M_M$ for every $M \in \max A_A$.

1.3.9 follows from 1.3.8(d),(e) and 1.3.6(b).

1.3.10 Invariant semiprime rings. Let A be a right invariant semiprime ring. Then A is a right localizable reduced ring.
The ring A is arithmetical if and only if A_M is a right uniserial domain for every $M \in \max A_A$.

1.3.11 The completion of the proof of Theorem 1C. Theorem 1C follows from 1.3.10.

2 Finitely generated modules and diagonalizability

The main results of this section are Theorems 2A and 2B.

2A Theorem (Golod [77]). If A is a commutative ring, then A is arithmetical if and only if $B + r(X) = r(X/XB)$ for every finitely generated A-module X and each ideal B of the ring A.

2B Theorem (Tuganbaev [187]). If A is a right invariant diagonalizable[1] ring, then $B + r(X) = r(X/XB)$ for every finitely generated right A-module X and each ideal B of the ring A.

Remark. It is clear that $B + r(X) \subseteq r(X/XB)$ for any right module X over an arbitrary ring A and each ideal B of the ring A.

Remark. For the completion of the proof of Theorems 2A and 2B, see 2.1.6 and 2.2.4.

2.1 Annihilators and finitely generated modules

2.1.1 Two remarks on arithmetical rings. Let A be an arithmetical ring and let B, $C_1, \dots, C_n$ be ideals of the ring A.

a. $B + C_1 \cap \dots \cap C_n = (B + C_1) \cap \dots \cap (B + C_n)$.

b. If $A_1, \dots, A_n$ are copies of the ring A and X is a right A-module $A_1/C_1 \oplus \dots \oplus A_n/C_n$, then $B + r(X) = r(X/(XB))$.

Proof. **a.** Since A is an arithmetical ring, the assertion is directly verified, with the use of the induction on n.

b. In our case, $B + r(X) = B + C_1 \cap \dots \cap C_n$ and $r(X/(XB)) = (B + C_1) \cap \dots \cap (B + C_n)$. Now we use **a.** □

2.1.2. Let A be a ring and let $(B + AcA) \cap (B + AdA) = B + (AcA) \cap (AdA)$ for every ideal B and any elements c, d of the ring A. Then A is an arithmetical ring.

Proof. Let B, C, D be ideals of the ring A and let $b_1 + c = b_2 + d \in (B+C) \cap (B+D)$, where $b_1, b_2 \in B$, $c \in C$ and $d \in D$. By assumption, $b_1 + c \in B + (AcA) \cap (AdA) \subseteq B + C \cap D$. Therefore, $(B + C) \cap (B + D) \subseteq B + C \cap D$. Consequently, A is an arithmetical ring. □

2.1.3 An arithmeticity criterion of invariant rings. If A is a right invariant ring, then the following conditions are equivalent.

1 The definition of a diagonalizable ring is given in 2.2.1.

https://doi.org/10.1515/9783110659825-002

1) A is an arithmetical ring.

2) $B + r(X) = r(X/(XB))$ for any right A-module X that is the finite direct sum of cyclic modules.

3) $B + r(X) = r(X/(XB))$ for any right A-module X that is the direct sum of two cyclic-presented modules.

Proof. The implication 2) ⇒ 3) is directly verified.

3) ⇒ 1). According to 2.1.2, it is sufficient to prove that $(B + AcA) \cap (B + AdA) = B + (AcA) \cap (AdA)$ for every ideal B and any elements c, d of the ring A. We denote by X the module $A_A/cA \oplus A_A/dA$, which is the direct sum of two cyclic-presented modules. Since A is a right invariant ring, $AcA = cA$ and $AdA = A$. It is directly verified that $B + r(X) = B + (AcA) \cap (AdA)$ and $(B + AcA) \cap (B + AdA) = r(X/(XB))$. In addition, $B + r(X) = r(X/(XB))$ by assumption. Therefore, $(B + AcA) \cap (B + AdA) = B + (AcA) \cap (AdA)$.

1) ⇒ 2). Since A is a right invariant ring, every cyclic right A-module is isomorphic to the module A_A/C for some two-sided ideal C of the ring A. Therefore, the assertion follows from 2.1.1(b). □

2.1.4. Let A be a ring and let $B+r(X) = r(X/XB)$ for every finite-presented right A-module X and each ideal B of the ring A.

a. $B + r(X) = r(X/(XB))$ for every finitely generated right A-module X and each ideal B of the ring A.

b. If the ring A is right invariant, then A is an arithmetical ring.

Proof. **a.** Let X be a finitely generated right A-module. Then $X \cong F/N$, where F is a finitely generated free A-module and N is some submodule in F. Let $\{N_i\}$ be the set of all finitely generated submodules in N. Then $N = \cup_i N_i$ and $M/MB \cong F/(N + FB) = F/\cup_i (N_i + FB)$. For each $i \in I$, we denote by X_i the finite-presented module F/N_i. By assumption, $B + r(X)_i = r(X_i/(X_iB)$ for every finite-presented module X_i. In addition, it is directly verified that $r(F/N) = \cup_i r(F/N_i)$. Therefore,

$$r(X/XB) = r(F/(N + FB)) = \cup_i r(F/(N_i + FB)) = \cup_i (B + r(F/N_i)) =$$
$$= B + \cup_i r(F/N_i) = B + r(F/N) = B + r(X)\,.$$

b. By **a**, $B + r(X) = r(X/(XB))$ for every finitely generated right A-module X and each ideal B of the ring A. By 2.1.3, A is an arithmetical ring. □

2.1.5 Theorem ([77]). A commutative ring A is arithmetical if and only if $B + r(X) = r(X/(XB))$ for every finitely generated A-module X and each ideal B of the ring A.

Proof. To prove that the relation $B+r(X) = r(X/(XB))$ is true for every finitely generated right A-module X and each ideal B of the ring A, it is sufficient to show that it is true for the localization by any maximal ideal of the ring A. Therefore, we can assume the ring A is local. By 1.2.2(e), any commutative arithmetical local ring is a uniserial. It is well known that every finite-presented module X over such a ring is the finite direct sum of cyclic modules, $X \cong A/B_1 \oplus \cdots \oplus A/B_n$, where we can assume that $B_1 \subseteq B_2 \subseteq \cdots \subseteq B_n$. Then

$$X/XB \cong A/(B + B_1) \oplus \cdots \oplus R/(B + B_n)\,,$$

whence $r(X/XB) = B + B_1 = B + r(X)$. □

2.1.6. If A is a right invariant, right Bezout ring and X is a finite direct sum of cyclic right A-modules, then $B + r(X) = r(X/(XB))$ for any ideal B of the ring A.

The assertion 2.1.6 follows from 2.1.3 and assertion III(c) from the introduction.

2.2 Diagonalizable rings

2.2.1 Diagonalizable matrices and rings. A rectangular matrix B over a ring A is said to be *diagonalizable* if there exist two square invertible matrices X and Y of suitable orders such that XBY is a diagonal matrix (i.e., all elements a_{ij} matrices XBY are equal to the zero for $i \neq j$).

A ring A is said to be *diagonalizable* if every rectangular matrix over A is diagonalizable. The ring of integers and commutative regular rings are examples of commutative diagonalizable rings.

Diagonalizable rings were studied in many papers. Here we only note the paper [106].

2.2.2 ([106, p. 477]). From the proof of [106, Theorem 9.1], we obtain the following well known assertion: every finite-presented right or left module over a diagonalizable ring is the finite direct sum of cyclic-presented modules.

2.2.3 Right (left) Hermite rings. A ring A is called a *right Hermite* ring if each row (a, b) of length 2, where $a, b \in A$, is diagonalizable, i.e., there exists an invertible 2×2 matrix Y such that $(a, b)Y = (c, 0)$ for some $c \in A$.
A ring A is said to be *left Hermite* if each column $\binom{a}{b}$ of height 2, where $a, b \in A$, is diagonalizable, i.e., there exists an invertible 2×2 matrix X such that $X\binom{a}{b} = \binom{c}{0}$ for some $c \in A$.

Hermite rings have been studied in many papers. Here we only indicate papers [9, 69, 70, 92, 106, 114, 193].

a. It is clear that every diagonalizable ring is a right and left Hermite ring.

b. Every right Hermite ring A is a right Bezout ring [106, p. 465].
Indeed, let $a, b \in A$. By assumption, there exists an invertible 2×2 matrix Y over

A that $(a, b)Y = (c, 0)$ for some $c \in A$. Then $c = ay_{11} + by_{21} \in aA + bA$. Since $(a, b) = (c, 0)Y^{-1}$, we have $a, b \in cA$.

c. There exists a commutative reduced Bezout ring A, which is not a Hermite ring. In particular, A is a nondiagonalizable ring [70, Example 3.4].

d. All right Bezout rings without zero-divisors are right Hermite; see [9] or 4.1.5, Section 4.

e. All commutative semihereditary Bezout rings are Hermite rings [114].

f. In [106, Theorem 3.5], it is proven that a ring A is a right Hermite ring if and only if for every rectangular $m \times n$ matrix M over A, there exists an invertible $m \times m$ matrix Y such that MY is a lower triangular matrix.
Similarly, a ring A is a right Hermite if and only if for every rectangular $m \times n$ matrix M over A, there exists an invertible $n \times n$ matrix X such that XM is a upper triangular matrix.

2.2.4 The completion of the proof of Theorems 2A and 2B. Theorem 2A follows from 2.1.5.

Proof. The proof of Theorem 2B. Let A be a right invariant diagonalizable ring. From 2.2.3(a),(b), A is a right invariant right Bezout ring. Then A is an arithmetical ring by the assertion III(c) from the introduction. Let B be an ideal of the ring A and let X be an arbitrary finite-presented right A-module. From 2.2.2, X is the finite direct sum of cyclic modules. From 2.1.3, $B + r(X) = r(X/XB)$. Since $B + r(X) = r(X/XB)$ for any finite-presented right A-module X, it follows from 2.1.4(a) that $B + r(Y) = r(Y/(YB))$ for any finitely generated right A-module Y. □

2.2.5 Open question. Let A be an invariant arithmetical ring and let X be an arbitrary direct summand of the finite direct sum of cyclic right A-modules. Is it true that $B + r(X) = r(X/(XB))$ for any ideal B of the ring A?

2.2.6 Open question. Let A be an invariant arithmetical ring and let X be an arbitrary finitely generated right A-module. Is it true that $B + r(X) = r(X/(XB))$ for any ideal B of the ring A?

3 Rings with flat and quasiprojective ideals

The main results of this section are Theorems 3A, 3B and 3C.

3A Theorem (Tuganbaev [157, 161, 182]). For an invariant semiprime ring A, the following conditions are equivalent.

1) A is an arithmetical ring.

2) Every submodule of any flat A-module is a flat module.

3) Every finitely generated ideal of the ring A is a quasiprojective right A-module.

3B Theorem (Jensen [98]). A commutative ring A is an arithmetical semiprime ring if and only if every submodule of any flat A-module is a flat module.

3C Theorem (Tuganbaev [182]). If A is an invariant ring, then A is an arithmetical ring if and only if every its finitely generated ideal is a quasiprojective right A-module such that all endomorphisms can be extended to endomorphisms of the module A_A.

Remark. The completion of the proof of Theorems 3A, 3B and 3C is given in 3.3.14.

3.1 Flat modules

3.1.1 Tensor products of modules and bimodules. Let A be a ring, X_A be a right A-module, ${}_AY$ be a left A-module, $X \times Y$ be the cartesian product, F be a free Abelian group with basis consisting of elements $X \times Y$, and let H be the subgroup in F generated by all elements of the form

$$(x + x', y) - (x, y) - (x', y)\,, \quad (x, y + y') - (x, y) - (x, y')\,, \quad (xa, y) - (x, ay)\,,$$
$$x, x' \in X\,, \quad y, y' \in Y\,, \quad a \in A\,.$$

The Abelian group F/H is called the *tensor product* of the modules X and Y; it is denoted by $X \otimes_A Y$. We write $X \otimes Y$ instead of $X \otimes_A Y$ if it is clear which ring A is meant. The image of the pair (x, y) under a natural mapping $X \times Y \to X \otimes Y$ is denoted by $x \otimes y$. If ${}_BX_A$, ${}_AY_C$ are two bimodules, then the group $X \otimes_A Y$ naturally turns into a B-C-bimodule such that the multiplications by $b \in B$ and $c \in C$ are defined by the relations $b(\sum x_iy_i) = \sum bx_iy_i$ and $(\sum x_iy_i)c = \sum x_iy_ic$, respectively. In particular, $X \otimes_A Y$ is an $\operatorname{End} X$-$\operatorname{End} Y$-bimodule and $X \otimes_A A$ is a right A-module.

If X is a right A-module and X' is a submodule in X such that a natural group homomorphism $X' \otimes_A Y \to X \otimes_A Y$ is a monomorphism, then we can consider $X' \otimes_A Y$ a subgroup in $X \otimes_A Y$ in this case.

https://doi.org/10.1515/9783110659825-003

The following properties **a–i** of tensor products are well known and are directly verified.

a. For any element x of the group $X \otimes_A Y$, there exists a finite set of subscripts I such that $x = \sum_{i \in I} x_i \otimes y_i$.

b. If $x, x' \in X$, $y, y' \in Y$ and $a \in A$, then

$$(x + x') \otimes y = x \otimes y + x' \otimes y\,, \quad x \otimes (y + y') = x \otimes y + x \otimes y'\,, \quad xa \otimes y = x \otimes ay\,.$$

c. If $\sum_{i=1}^{n} x_i \otimes y_i \in X \otimes_A Y$, then $\sum_{i=1}^{n} x_i \otimes y_i = 0$ if and only if there exist finite sets $\{\overline{x}_k \in X\}_{k=1}^{m}$ and $\{a_{ik} \in A\}$ such that

$$1 \le i \le n\,, \quad 1 \le k \le m\,, \quad x_i = \sum_{k=1}^{m} \overline{x}_k a_{ik}\,, \quad \sum_{i=1}^{n} a_{ik} y_i = 0 \quad \forall i, k\,.$$

d. For any two module homomorphisms $f\colon X_A \to M_A$ and $g\colon {}_AY \to {}_AN$, the relation $(f \otimes g)(\sum x_i \otimes y_i) = \sum f(x_i) \otimes g(y_i)$ correctly defines the group homomorphism $f \otimes g\colon X \otimes {}_AY \to M \otimes {}_AN$.

e. The canonical group epimorphism $h\colon X \otimes_A A \to X$ is an isomorphism from a natural right A-module $X \otimes {}_AA$ onto the module X_A.

f. If $X_A = \oplus_{i \in I} X_i$, then there exists a natural group isomorphism from the group $(\oplus_{i \in I} X_i) \otimes {}_AY$ on to the group $\oplus_{i \in I}(X_i \otimes {}_AY)$.

g. If P, Q are two submodules in ${}_AY$, then the intersection in $X \otimes {}_A(P + Q)$ of canonical images of the modules $X \otimes {}_AP$ and $X \otimes {}_AQ$ coincides with canonical image $X \otimes {}_A(P \cap Q)$.

h. If ${}_BX_A$, ${}_AY_C$ and ${}_CZ_D$ are three bimodules, then there exists a natural B-D-bimodule isomorphism ${}_B((X \otimes_A Y) \otimes_C Z)_D \to {}_B(X \otimes_A (Y \otimes_C Z))_D$.

i. If ${}_BX_A$, ${}_AY_C$, ${}_DZ_C$ are three bimodules and ${}_BX'_A$ is a subbimodule in ${}_BX_A$, then there exist natural B-D-bimodule isomorphisms

$$\alpha\colon \operatorname{Hom}((X \otimes_A Y)_C, Z_C) \to \operatorname{Hom}(X_A, \operatorname{Hom}(Y_C, Z_C))\,,$$
$$\beta\colon \operatorname{Hom}((X' \otimes_A Y)_C, Z_C) \to \operatorname{Hom}(X'_A, \operatorname{Hom}(Y_C, Z_C))$$

and natural B-D-bimodule homomorphisms

$$f\colon \operatorname{Hom}((X \otimes_A Y)_C, Z_C) \to \operatorname{Hom}((X' \otimes_A Y)_C, Z_C)\,,$$
$$g\colon \operatorname{Hom}(X_A, \operatorname{Hom}(Y_C, Z_C)) \to \operatorname{Hom}(X'_A, \operatorname{Hom}(Y_C, Z_C))$$

with $g\alpha = \beta f$. Therefore, if f is an epimorphism, then g is an epimorphism.

3.1.2 Flat modules. The first criterion. A right module X over a ring A is said to be *flat* if for any monomorphism of left A-modules $Y' \to Y$, a natural group homomorphism $X \otimes {}_AY' \to X \otimes {}_AY$ is a monomorphism.

For a ring A and a right A-module X, the following conditions are equivalent.

1) X is a flat module.

2) For every left ideal Y of A, the canonical group epimorphism $X \otimes_A Y \to XY$ is an isomorphism.

3) For every finitely generated left ideal Y of A, the canonical group epimorphism $X \otimes_A Y \to XY$ is an isomorphism.

4) For any elements $x_1, \ldots, x_n \in X$ and $y_1, \ldots, y_n \in A$ with $\sum_{i=1}^n x_i y_i = 0$, there exist elements $\overline{x}_1, \ldots, \overline{x}_m \in X$ and $a_{ik} \in A$ such that $1 \le i \le n$, $1 \le k \le m$, $x_i = \sum_{k=1}^m \overline{x}_k a_{ik}$ and $\sum_{i=1}^n a_{ik} y_i = 0$ for all i and k.

Proof. The equivalence of conditions 1), 2) and 3) is verified with the use of 3.1.1(c).

2) $\Rightarrow$ 4). We set $Y = \sum_{i=1}^n A y_i \subseteq A$. Since $\sum_{i=1}^n x_i y_i = 0$, it follows from 2) that $\sum_{i=1}^n x_i \otimes y_i = 0$. Now we use 3.1.1(c).

4) $\Rightarrow$ 2). Let Y be a left ideal in A, $h\colon X \otimes Y \to XY$ be the canonical group epimorphism, and let $\sum_{i=1}^n x_i \otimes y_i \in \operatorname{Ker} h$. Then $\sum_{i=1}^n x_i y_i = 0$ and there exist elements $\overline{x}_1, \ldots, \overline{x}_m \in X$ and $a_{ik} \in A$ such that

$$1 \le i \le n\,, \quad 1 \le k \le m\,, \quad x_i = \sum_{k=1}^m \overline{x}_k a_{ik}\,, \quad \sum_{i=1}^n a_{ik} y_i = 0\,,$$

$$\sum_{i=1}^n x_i \otimes y_i = \sum_{i=1}^n \sum_{k=1}^m \overline{x}_i a_{ik} \otimes y_i = \sum_{k=1}^m \sum_{i=1}^n \overline{x}_i \otimes a_{ik} y_i = \sum_{i=1}^n \overline{x}_i \otimes 0 = 0\,.$$

Therefore, h is an isomorphism. □

3.1.3 Properties of flat modules. **a.** All direct sums of and all direct summands of flat modules are flat.

b. All free or projective modules are flat.

Proof. **a.** The assertion follows from 3.1.1(f).

b. By 3.1.1(e) and 3.1.2, all free cyclic modules are flat. Since every free module is a direct sum of free cyclic modules, it follows from **a** that all free modules are flat. Since every projective module is isomorphic to the direct summand of a free module, it follows from **a** that all projective modules are flat. □

3.1.4 Hattori torsion-free modules. For a ring A, a right A-module X is said to be *Hattori torsion-free* or *H-torsion-free* if for any $a \in A$, a natural group epimorphism $X \otimes Aa \to Xa$ is an isomorphism; see [89].

For a right module X over the ring A, the following conditions are equivalent.

1) X is an H-torsion-free module.

2) In X, every cyclic submodule is contained in some submodule in X, which is an H-torsion-free module.

3) For any elements $x \in X$ and $a \in A$ with $xa = 0$, there exist elements $x_1, \dots, x_n \in X$ and $a_1, \dots, a_n \in A$ such that $x = \sum_{i=1}^{n} x_i a_i$ and $a_i a = 0$ for all i.

Proof. The equivalence 1) ⇔ 2) is directly verified.

1) ⇒ 3). We assume that $xa = 0$, where $x \in X$ and $a \in A$. By assumption, $x \otimes a = 0$. By 3.1.1(c), $x = \sum_{j=1}^{k} x_j a_j$, where $x_j \in X$, $a_j \in A$ and $a_j a = 0$ for all j. Therefore, X is H-torsion-free.

3) ⇒ 1). We assume that $\sum_{i=1}^{n} x_i b_i a = 0$, where $x_i \in X$ and $b_i \in A$. We set $x = \sum_{i=1}^{n} x_i b_i \in X$. Then $xa = 0$. Since X is an H-torsion-free module, $x = \sum_{j=1}^{k} x_j a_j$, where $x_j \in X$, $a_j \in A$ and $a_j a = 0$ for all j. Then

$$\sum_{i=1}^{n} x_i \otimes b_i a = x \otimes a = \sum_{j=1}^{k} x_j a_j \otimes a = \sum_{j=1}^{k} x_j \otimes a_j a = \sum_{j=1}^{k} x_j \otimes 0 = 0\,. \quad \square$$

□

3.1.5 Properties of Hattori torsion-free modules. a. Every flat module is an H-torsion-free module.

b. If X' is a submodule of the right module X over the ring A, then X/X' is an H-torsion-free module if and only if
for any elements $x \in X$ and $a \in A$ with $xa \in X'$, there exist elements $x_1, \dots, x_n \in X$ and $a_1, \dots, a_n \in A$ such that $x = \sum_{i=1}^{n} x_i a_i \in X'$ and $a_i a = 0$ for all i.

c. If a is an element of the ring A, then $(A/aA)_A$ is an H-torsion-free module if and only if the right ideal aA is generated by an idempotent, i.e., A/aA is a projective A-module.

Proof. **a, b.** The assertions are verified with the use of 3.1.4.

c. If A/aA is a projective A-module, it follows from 3.1.3(b) that A/aA is an H-torsion-free module.

We assume that A/aA is an H-torsion-free module. Since $1 \cdot a \in aA$ and A/aA is an H-torsion-free right A-module, it follows from **b** that there exist elements $b, x_1, \dots, x_n, a_1, \dots, a_n \in A$ such that $1 = ab + \sum_{i=1}^{n} x_i a_i$ and $a_i a = 0$ for all i. Then $a = (ab + \sum_{i=1}^{n} x_i a_i)a = aba$. Therefore, ab is an idempotent and $abA = aA$. □

3.1.6 Flat modules. The second criterion ([93]). For a ring A and a right A-module X, the following conditions are equivalent.

1) X is a flat module.

2) X is an H-torsion-free module and $XB \cap XC = X(B \cap C)$ for any left ideals B and C in A.

3) X is an H-torsion-free module and $XB \cap XC = X(B \cap C)$ for any finitely generated left ideals B and C in A.

4) In the module X, every finitely generated submodule is contained in some flat submodule of X.

Proof. The implications 2) ⇒ 3) and 1) ⇒ 4) are obvious.

3) ⇒ 1). Let $k \in \mathbb{N}$ and let Y be a k-generated left ideal. From 3.1.2, it is sufficient to prove that the canonical group epimorphism $f_Y : X \otimes Y \to XY$ is an isomorphism.

We use induction on k. For $k = 1$, the assertion follows from the property that M is an H-torsion-free module. Let $Y = B + C$ be a k-generated left ideal, where B is a $(k-1)$-generated left ideal and C is a principal left ideal. Let

$$f_B : X \otimes B \to XB\,, \quad f_C : X \otimes C \to XC\,, \quad w : X \otimes A \to XA\,,$$
$$h_B : X \otimes B \to X \otimes Y\,, \quad h_C : X \otimes C \to X \otimes Y\,, \quad g : X \otimes Y \to X \otimes A$$

be canonical group homomorphisms and let

$$p = \sum x_i \otimes (b_i - c_i) = \sum x_i \otimes b_i - \sum x_i \otimes c_i \in \operatorname{Ker} f_Y\,,$$
$$b_i \in B\,, \quad c_i \in C\,, \quad q \equiv \sum x_i c_i \in XB\,, \quad s \in X \otimes B\,, \quad \sum x_i \otimes b_i = h_B(s)\,.$$

Since $0 = f_Y(p) = \sum x_i(b_i - c_i)$, we have $q = \sum x_i c_i \in XB \cap u_j \otimes d_j = h_B(v)$. Then

$$0 = f_Y \left(\sum u_j \otimes d_j - \sum x_i \otimes b_i \right) = wg(h_B(v - s)) = f_B(v - s)\,,$$

whence $v = s$, since it follows from the induction hypothesis that f_B is a monomorphism. Then

$$\sum x_i \otimes b_i = h_B(s) = h_B(v) = \sum u_j \otimes d_j\,.$$

Similarly, $\sum x_i \otimes c_i = \sum u_j \otimes d_j$. Therefore, $p = \sum x_i \otimes b_i - \sum x_i \otimes c_i = 0$ and f is an isomorphism.

1) ⇒ 2). By 3.1.1(g), the intersection in $X \otimes_A (B + C)$ of canonical images of the modules $X \otimes_A B$ and $X \otimes_A C$ coincides with canonical image $X \otimes_A (B \cap C)$. By applying 3.1.2, we obtain the required assertion.

4) ⇒ 3). From 3.1.4, X is an H-torsion-free module. Let B, C be two finitely generated left ideals and

$$x = \sum_{i=1}^{n} y_i b_i = \sum_{j=1}^{m} z_j c_j \in XB \cap XC\,, \quad y_i, z_j \in X\,, \quad b_i \in B\,, \quad c_j \in C\,.$$

We denote by Y the submodule in X generated by all elements y_i and z_j. Since Y is finitely generated, it follows from the assumption that Y is contained in some flat submodule M in X. Therefore, $x \in MB \cap MC$. Since M is a flat module and the implication 1) ⇒ 2) has been proven, $MB \cap MC = M(B \cap C) \subseteq X(B \cap C)$. Therefore, $x \in X(B \cap C)$ and $XB \cap XC = X(B \cap C)$. □

3.1.7 Exact sequences of modules. A sequence of module homomorphisms $\cdots \longrightarrow M_{n-1} \xrightarrow{f_{n-1}} M_n \xrightarrow{f_n} M_{n+1} \xrightarrow{f_{n+1}} \ldots$ is said to be *exact in the term* M_n if $f(M_{n-1}) = \operatorname{Ker} f_n$ for all n.

A sequence of homomorphisms is said to be *exact* if it is exact in every its term.

With the use of 3.1.1, it is directly verified that
if $0 \longrightarrow X_1 \xrightarrow{f} X_2 \xrightarrow{g} X_3 \longrightarrow 0$ is an exact sequence of right A-modules, then $X_1 \otimes Y \xrightarrow{f\otimes 1} X_2 \otimes Y \xrightarrow{g\otimes 1} X_3 \otimes Y \longrightarrow 0$ is an exact sequence of Abelian groups for any left A-module Y.
In particular, $g \otimes 1$ is an epimorphism and $X_3 \otimes Y \cong (X_2 \otimes Y)/(f \otimes 1)(X_1 \otimes Y)$.

3.1.8 Flat modules. The third criterion. Let A be a ring, P be a flat right A-module, and let Q be a submodule in P. The following conditions 1)–5) are equivalent.

1) P/Q is a flat module.

2) For any finitely generated left ideal B of A, a natural group epimorphism $PB/QB \to (P/Q)B$ is an isomorphism.

3) For any left ideal B of A, a natural group epimorphism $PB/QB \to (P/Q)B$ is an isomorphism.

4) $Q \cap PB = QB$ for any finitely generated left ideal B of the ring A.

5) $Q \cap PB = QB$ for any left ideal B in A.

Under the conditions 1)–5), Q is a flat module.

Proof. According to 3.1.7, the sequence $Q \otimes B \longrightarrow P \otimes B \longrightarrow (P/Q) \otimes M \longrightarrow 0$ is exact. Since P is a flat module, it follows from 3.1.2 that the canonical epimorphism $P \otimes B \to PB$ is an isomorphism that maps from $Q \otimes B$ onto QB. Therefore, $(P/Q)B \cong PB/QB$. From 3.1.2, P/Q is a flat module if and only if the canonical epimorphism $(P/Q) \otimes B \to (P/Q)B$ is an isomorphism for any (finitely generated) left ideal B.
The equivalence of conditions 1), 2) and 3) follows from the above.

The equivalence of conditions 1), 4) and 5) follows from the equivalence conditions 1), 2), 3) and the property that $(P/Q)B = (PB + Q)/Q \cong PB/(Q \cap PB)$.

5) $\Rightarrow$ 1). We have to prove that Q is a flat module. Since P is a flat module, it follows from 3.1.6 that P is an H-torsion-free module and $PB \cap PC = P(B \cap C)$. Therefore, it follows from 5) that

$$QB \cap QC = Q \cap PB \cap PC = Q \cap P(B \cap C) = Q(B \cap C) .$$

From 3.1.6, it is sufficient to prove that Q is an H-torsion-free module. Let $qa = 0$, where $q \in Q$ and $a \in A$. Since P is an H-torsion-free module, there exist elements $p_1, \dots, p_n \in P$ and $d_1, \dots, d_n \in A$ such that $q = \sum_{i=1}^n p_i d_i$ and $d_i a = 0$ for all i. Let $D = \sum_{i=1}^n Ad_i$. Then $q \in Q \cap PD$. Therefore, $q \in QD$ by 5). Then there exist elements $q_1, \dots, q_n \in Q$ with $q = \sum_{i=1}^n q_i d_i$. Since $d_i a = 0$ for all i, we have that Q is an H-torsion-free module. □

3.1.9 Flat modules. The fourth criterion. Let A be a ring, X be a free right A-module, and let Y be a submodule of X. The following conditions are equivalent.

1) X/Y is a flat module.

2) For any $y \in Y$, there exists a homomorphism $h\colon X \to Y$ with $h(y) = y$.

3) For any finite subset $\{y_1, \dots, y_n\} \subseteq Y$, there exists a homomorphism $h\colon X \to Y$ such that $h(y_i) = y_i$ for all $y_1, \dots, y_n$.

Proof. 1) $\Rightarrow$ 2). Let $\{x_i\}_{i\in I}$ be a basis of the free module X_A, $\{x_1, \dots, x_n\} \subseteq \{x_i\}_{i\in I}$, and let $y = \sum_{i=1}^n x_i a_i$, where $a_1, \dots, a_n \in A$. We set $B = \sum_{i=1}^n Aa_i$. We have $y \in Y \cap XB$ and $Y \cap XB = YB$, according to 3.1.8. Therefore, $y = \sum_{i=1}^k y_i b_i$ for some elements $y_1, \dots, y_k \in Y$ and $b_1, \dots, b_k \in B$. There exists a homomorphism $h\colon X \to X$ such that $h(x_i) = y_i$ for all $i = 1, \dots, k$ and $h(x_i) = 0$ for all $i \in I \setminus \{1, \dots, k\}$. Then $h(X) \subseteq Y$ and $h(y) = y$.

2) $\Rightarrow$ 3). We use induction on n. For $n = 1$, the assertion follows from 2). We assume that $n > 1$. Let $f\colon X \to Y$ be a homomorphism such that $f(y_n) = y_n$. By the induction hypothesis, there exists a homomorphism $\varphi\colon X \to Y$ such that $\varphi(y_i - f(y_i)) = y_i - f(y_i)$ for all $i = 1, \dots, n-1$. Therefore, $(1_X - \varphi)(1_X - f)(y_n) = (1_X - \varphi)(y_n - y_n) = 0$. For any $i < n$, we obtain that $(1_X - \varphi)(1_X - f)(y_i) = (1_X - \varphi)(y_i - f(y_i)) = 0$. We set $h = 1_X - (1_X - \varphi)(1_X - f) \in \mathrm{End}(X)$. Then h has the required properties.

The implication 3) $\Rightarrow$ 2) is obvious.

2) $\Rightarrow$ 1). Let B be a finitely generated right ideal in A and $y = \sum_{i=1}^n x_i b_i \in Y \cap XB$, where $x_i \in X$ and $b_i \in B$. By assumption, there exists a homomorphism $h\colon X \to Y$ such that $y = h(y)$. Then $y = h(y) = \sum_{i=1}^n h(x_i) b_i \in YB$, whence $Y \cap XB = YB$. From 3.1.8, X/Y is a flat module. □

3.1.10. If M is a finitely-presented flat module, then M is a projective module.

Proof. Let $M \cong F/Q$, where F is a finitely generated free module and Q is generated by elements $q_1, \dots, q_n$. By 3.1.9, there exists a homomorphism $h\colon F \to Q$ such that $h(q_i) = q_i$ for all i. Then $h(q) = q$ for any $q \in Q$. Therefore, M is isomorphic to a direct summand of the free module F. □

3.1.11 Modules of fractions. Let A be ring, S be a right denominator set in A, Q be the right ring of fractions of A with respect to S, $f\colon A \to Q$ be the canonical ring homomorphism, and let M be a right A-module.

For the cartesian product $M \times S$, we define a relation $\sim$ such that

$$(x, s) \sim (y, t) \quad \Leftrightarrow \quad \exists c, d \in A\colon\ xc = yd\,,\ sc = td \in S\,.$$

It is easy to verify that $\sim$ is an equivalence relation for $M \times S$.

Let h be a natural surjective mapping $M \times S \to (M \times S)/\sim\ \equiv MS^{-1}$ and let g_S be a mapping from M into MS^{-1} such that $g_S(m) = h(m, 1)$. In MS^{-1}, we define an addition and a multiplication by elements of Q using the relations $h((x, s)) + h((y, t)) = h((xc +$

$yd, u))$, $u = sc = td \in S$ and $h((x, c)) \cdot f(b)f(t)^{-1} = h((xc, tu))$, where $b \in A$, $t \in S$, $sc = bu$ and $u \in S$.

It is directly verified that MS^{-1} is a right Q-module, $g_S \equiv g$ is an A-module homomorphism, $\operatorname{Ker} g = \{m \in M \mid \exists s \in S\colon ms = 0\}$ and $MS^{-1} = \{g(m)f(s)^{-1} \mid m \in M, s \in S\}$.

The right Q-module MS^{-1} is called the *module of fractions* of M with respect to S. The A-module homomorphism $g_S\colon M \to MS^{-1}$ is called the *canonical* homomorphism.
For each submodule X in A-module M, we denote by $\overline{X}$ and XS^{-1} a natural image of the module X in A-module $M/\operatorname{Ker} g$ and the submodule $g(X)Q$ in Q-module MS^{-1}.

a. The rule $\varphi(\sum_{i=1} m_i \otimes q_i) = \sum_{i=1} g(m_i)q_i$ correctly defines an isomorphism of right Q-modules $\varphi\colon M \otimes_A AS^{-1} \to MS^{-1}$.

b. For any $x_1, \ldots, x_n \in MS^{-1}$, there exist elements $s \in S$ and $m_1, \ldots, m_n \in M$ such that $x_i = g(m_i)f(s)^{-1}$ for $i = 1, \ldots, n$.

c. If X is a submodule in M and $\overline{X}$ is the S-saturation of the module X in M, then

$$XS^{-1} = \overline{X}S^{-1} = \{g(x)f(t)^{-1} \mid x \in X\,,\ t \in S\}\,.$$

Therefore, for any $y \in XS^{-1}$ there exists an element $s \in S$ with $yf(s) \in g(X)$.

d. If W is a submodule in the Q-module MS^{-1} and $X = g^{-1}(W \cap g(M))$ is a submodule of an A-module M, then $XS^{-1} = W$, $X \supseteq \operatorname{Ker} g$, and X is an S-saturated submodule in M.
The correspondence $X \to XS^{-1}$ is an inclusion-preserving bijection between the set of all S-saturated submodules A-module M and the set of all submodules Q-module MS^{-1}.

e. Let X, Y be two submodules in M_A. Then $(X+Y)S^{-1} = XS^{-1} + YS^{-1}$ and $(X \cap Y)S^{-1} \subseteq XS^{-1} \cap YS^{-1}$.
In addition, if $\operatorname{Ker} g \subseteq X \cap Y$, then $(X \cap Y)S^{-1} = XS^{-1} \cap YS^{-1}$.
Therefore, the mapping of $\overline{X} \to XS^{-1}$ is a surjective lattice homomorphism from the lattice of all submodules A-module $M/\operatorname{Ker} g$ onto the lattice of all submodules Q-module MS^{-1}.
Consequently, if M is a distributive (resp., Noetherian, Artinian) A-module, then MS^{-1} is a distributive (resp., Noetherian, Artinian) Q-module.

f. If t is a cardinal number and X is a t-generated submodule in M_A, then XS^{-1} is a t-generated submodule of the Q-module MS^{-1}.
Consequently, if all submodules of an A-module M are t-generated, then all submodules of the Q-module MS^{-1} are t-generated.
In particular, if M is a Bezout A-module, then MS^{-1} is a Q-Bezout module.
In addition, for any t-generated submodule P_Q in $(MS^{-1})_Q$, there exists a t-generated submodule X in M_A with $XS^{-1} = P$.

g. If $P_Q = \oplus_{i=1}^{n} P_i$ is a submodule of the Q-module MS^{-1}, then the submodule $P \cap g(M)$ of the module $g(M)_{f(A)}$ is the direct sum of the modules $P_i \cap g(M)$.

h. ${}_AQ$ is a flat module.

i. $(X/Y)S^{-1} \cong XS^{-1}/YS^{-1}$ for any submodule X of M and any submodule Y of X.

j. If $m \in M$ and X is a submodule in M, then $g(m) \in g(X)Q$ if and only if $ms \in X$ for some $s \in S$.

k. If M_A is a free (resp., projective, flat) module, then $(MS^{-1})_Q$ is a free (resp., projective, flat) module.

l. If $m \in M$ and X is a submodule in M, then $f((m^{\cdot}.X))Q \subseteq (g(mA)Q^{\cdot}.g(X)Q)$.

m. If t is a cardinal number and all t-generated right ideals of the ring A are free (resp., projective, flat) A-modules, then all t-generated right ideals of the ring Q are free (resp., projective, flat) Q-modules.
In particular, if the ring A is right semihereditary, then the ring Q is right semihereditary.

Proof. **a.** The assertion is directly verified.

b. By definition of the module of fractions MS^{-1}, there exist elements $s_1, \dots, s_n \in S$ and $y_1, \dots, y_n \in M$ such that $x_i = g(y_i)f(s_i)^{-1}$ for $i = 1, \dots, n$. According to 1.3.5(c), there exist elements $s \in S$ and $a_1, \dots, a_n \in A$ such that $f(s_i)^{-1} = f(a_i)f(s)^{-1}$ for $i = 1, \dots, n$. If we set $m_i = y_i a_i$, then

$$x_i = g(y_i)f(s_i)^{-1} = g(y_i)f(a_i)f(s)^{-1} = g(m_i)f(s)^{-1}, \quad i = 1, \dots, n.$$

c. The assertion follows from **b** and 1.2.1(c).

d. The assertion follows from **b, c** and 1.2.1(c).

e. It is clear that $(X \cap Y)S^{-1} \subseteq XS^{-1} \cap YS^{-1}$. Let $w_1 \in XS^{-1} + YS^{-1}$. According to **b**, there exist elements $x_1 \in X$, $y_1 \in Y$ and t such that

$$w_1 = g(x_1)f(t)^{-1} + g(y_1)f(t)^{-1} = g(x_1 + y_1)f(t)^{-1} \in (X + Y)S^{-1}.$$

Therefore, $XS^{-1} + YS^{-1} \subseteq (X+Y)S^{-1} \subseteq XS^{-1} + YS^{-1}$. We assume that $\operatorname{Ker} g \subseteq X \cap Y$ and $w_2 \in XS^{-1} \cap YS^{-1}$. According to **b**, $w_2 = g(x_2)f(u)^{-1} = g(y_2)f(u)^{-1}$ for some $x_2 \in X$, $y_2 \in Y$ and $u \in S$. Then $g(x_2) = g(y_2)$, $x_2 - y_2 \in \operatorname{Ker} g \subseteq X \cap Y$, and $x_2 v = y_2 v \in X \cap Y$ for some $v \in S$. Then $w_2 = g(x_2 v)f(u)^{-1}f(t)^{-1} \in (X \cap Y)S^{-1}$.

f. The assertion is verified with the use of **b** and **c**.

g. The assertion is verified with the use of **e**.

h. Let $h\colon X_A \to M_A$ be a monomorphism of right A-modules, $x \in X$ and $g(h(x)) = 0$. Then $h(x)s = 0$ for some $s \in S$, whence $xs \in \operatorname{Ker} h = 0$ and $(xA)S^{-1} = 0$, which is required.

i. The assertion follows from **h** and the property that $(XS^{-1})_B \cong (X \otimes_A Q)_Q$ for any module X_A.

j. If $ms \in X$ for some $s \in S$, then $g(m)f(s) \in g(X)$ and $g(m) \in g(X)f(s)^{-q} \subseteq g(X)Q$. Now we assume that $g(m) \in g(X)Q$. According to **c**, $g(m) = g(x)f(t)^{-1}$ for some $x \in X$ and

$t \in S$. Then $g(mt) = g(x)$. Since $mt - x \in \operatorname{Ker} g$, we have that $(mt - x)u = 0$ for some $u \in S$. We set $s = tu \in S$. Then $ms = mtu = xu \in X$.

k. Since Q_Q is free, the assertion follows from 3.1.1(h) and the relation $MS^{-1} = M \otimes_A Q$.

l. Let $y \in (g(m)Q\,^{\cdot}.\,g(X)Q)$. Then $y = f(a)f(s)^{-1}$ for some $a \in A$ and $s \in S$. Then $g(ma) = g(m)f(a)f(s)^{-1}f(s) \in g(m)yQ \subseteq g(X)Q$. According to **j**, $mat \in X$ for some $t \in S$. Then $at \in (m\,^{\cdot}.\,X)$ and $y = f(at)f(t)^{-1}f(s)^{-1} \in f((m\,^{\cdot}.\,X))Q$.

m. The assertion follows from **k** and **f**. □

3.2 Flat ideals and submodules

3.2.1. In a module X, all submodules are H-torsion-free modules if and only if all cyclic submodules of X are H-torsion-free modules.
All submodules of X are flat modules if and only if all finitely generated submodules in X are flat.

The first assertion follows from 3.1.4, and the second follows from 3.1.6.

3.2.2 Theorem on flat submodules. For a ring A, the following conditions are equivalent.

1) All submodules of flat right A-modules are flat.

2) All submodules of flat left A-modules are flat.

3) In A, all right ideals are flat.

4) In A, all left ideals are flat.

5) In A all finitely generated right ideals are flat.

6) In A, all finitely generated left ideals are flat.

7) For any finitely generated right ideal X and each finitely generated left ideal Y of A, a natural group homomorphism $X \otimes Y \to XY$ is an isomorphism.

Proof. Since condition 7) is left-right symmetrical, it is sufficient to prove the equivalence of conditions 1), 3), 5) and 7).

The implication 1) ⇒ 3) follows from the property that A_A is a flat module according to 3.1.3(b).

The equivalence 3) ⇔ 5) follows from 3.2.1.

The equivalence 5) ⇔ 7) follows from 3.1.2.

7) ⇒ 1). Let X be a submodule of the flat module M_A, $f\colon X \to M$ be a natural embedding, and let Y be a finitely generated left ideal of A. By the left-right symmetrical analog of 3.1.2, ${}_AY$ is a flat module. Therefore, $f \otimes 1$ is a monomorphism. We assume that $\sum_{i=1}^{n} x_i y_i = 0$, where $x_1, \ldots, x_n \in X$ and $y_1, \ldots, y_n \in Y$. Since M is a flat module, it

follows from 3.1.2 that a natural group epimorphism $M \otimes Y \to MY$ is an isomorphism. Therefore, $(f \otimes 1)(\sum_{i=1}^{n} x_i \otimes y_i) = 0$. Since $f \otimes 1$ is a monomorphism, $\sum_{i=1}^{n} x_i y_i = 0$. Therefore, a natural group epimorphism $X \otimes Y \to XY$ is an isomorphism. According to 3.1.2, X is a flat module. □

3.2.3 Flat principal right ideals and PF rings. A ring A is called a *PF ring* if for any two elements $x, y \in A$ with $xy = 0$, there exist elements $a, b \in A$ such that $a+b = 1$, $xa = 0$ and $by = 0$.

a. Let A be a ring and x an element of A. The principal right ideal xA is a flat A-module if and only if for any $y \in A$ with $xy = 0$, there exist two elements $a, b \in A$ such that $a + b = 1$, $xa = 0$ and $by = 0$.

b. A ring A is a PF ring if and only if all its principal right ideals are flat right A-modules.

c. A ring A is a PF ring if and only if all its principal left ideals are flat left A-modules.

d. If A is a unitary subring of a local ring Q, then A is a PF ring if and only if A is a domain.

e. If A is a PF ring and S is a right Ore set in A, then the right ring of fractions AS^{-1} is a PF ring.

f. If A is a commutative PF ring, then A is a reduced ring.

Proof. **a.** Since $xA \cong A_A/r(x)$ and $\mathrm{Hom}(A_A, r(x))$ can be identified with $r(x)$, it follows from 3.1.9 that xA is a flat module if and only if for any element $y \in r(x)$, there exists an element $a \in r(x)$ with $y = ay$. This implies **a**.

b. The assertion follows from **a**.

c. The assertion follows from **b** and the property that the property to be PF ring is left-right symmetrical.

d. If A is a domain, then all its nonzero principal right ideals are free and A is a PF ring.

We assume that A is a PF ring, $x, y \in A$ and $xy = 0$. According to 3.2.3, there exist two elements $a, b \in A$ such that $a + b = 1$, $xa = 0$ and $by = 0$. Since $a + b = 1$ and a, b are elements of the local ring Q, at least one of the elements a, b is invertible in Q; therefore, the element is a nonzero divisor in A. If a is a nonzero divisor, then $x = 0$, since $xa = 0$. If b is a nonzero divisor, then $y = 0$, since $by = 0$.

e. Let $p, q \in AS^{-1}$ and $pq = 0$. Per 1.3.2(a), we can assume that $p = xs^{-1}$ and $q = zs^{-1}$, where $x, z \in A$ and $s \in S$. Then $xs^{-1}z = 0$. There exist elements $y \in A$ and $t \in S$ such that $s^{-1}z = yt^{-1}$. Then $xy = 0$. Since A is a PF ring, there exist two elements $a, b \in A$ such that $a + b = 1$, $xa = 0$ and $by = 0$. Therefore, $ps = xs^{-1}sa = 0$, whence $p(sas^{-1}) = 0$. In addition, $0 = byt^{-1} = bs^{-1}z$, whence $(sbs^{-1})q = s(bs^{-1}z)s^{-1} = 0$. Since $sas^{-1} + sbs^{-1} = 1$, we have that AS^{-1} is a PF ring.

f. Let $x \in A$ and $x^2 = 0$. Since A is a commutative PF ring, there exist elements $a, b \in A$ such that $a + b = 1$, $xa = 0$ and $bx = xb = 0$. Therefore, $x = x(a + b) = 0$. □

3.2.4. Let A be a reduced ring and $x \in A$. The following conditions are equivalent.

1) xA_A is a flat module.

2) For any element $y \in A$ with $(AxA) \cap (AyA) = 0$, there exist two elements $a, b \in A$ such that $a + b = 1$ and $xa = yb = 0$.

3) $xA \oplus yA = (x + y)A$ for any element $y \in A$ with $AxA \cap AyA = 0$.

4) $xA \oplus yA$ is a principal right ideal for any element $y \in A$ with $AxA \cap AyA = 0$.

Proof. The equivalence 1) ⇔ 2) follows from 3.2.3(a).

2) ⇒ 3). It is sufficient to prove that $x \in (x+y)A$ and $y \in (x+y)A$. According to 2), there exist elements $a, b \in A$ such that $a + b = 1$ and $xa = yb = 0$. Then

$$x = x(a + b) = xb = (x + y)b \in (x + y)A\,,$$
$$y = y(a + b) = ya = (x + y)a \in (x + y)A\,.$$

3) ⇒ 4). The assertion is directly verified.

4) ⇒ 2). By 4), $xA \oplus yA = zA$. There exist elements $f, g \in A$ such that $z = xf + yg$. Then there exist elements $u, v \in A$ such that $x = (xf + yg)u$ and $y = (xf + yg)v$. Then

$$x(1 - fu) = ygu \in xA \cap yA = 0\,,$$
$$y(1 - gv) = xfv \in xA \cap yA = 0\,.$$

Then $x = xfu$ and $v \in r(xf)$. We set $b = 1 - gv$ and $a = 1 - b = gv$. Then $yb = 0$. Since $v \in r(xf)$, it follows from 1.3.1(d) that $ugv \in r(xf)$. Therefore, $xa = xfua = xfugv = 0$. □

3.2.5 PF rings without nilpotent elements. Let A be a reduced ring. The following conditions are equivalent.

1) A is a PF ring.

2) For any elements $x, y \in A$ with $AxA \cap AyA = 0$, there exist two elements $a, b \in A$ such that $a + b = 1$, $AxA \cap AaA = 0$ and $AbA \cap AyA = 0$.

3) For any elements $x, y \in A$ with $AxA \cap AyA = 0$, there exist elements two elements $a, b \in A$ such that $a + b = 1$, $xa = 0$ and $yb = 0$.

4) The 2-generated right ideal $xA \oplus yA$ coincides with the principal right ideal $(x + y)A$ for any two elements $x, y \in A$ with $AxA \cap AyA = 0$.

5) $(x+y)A = (x+y)A \cap xA \oplus (x+y)A \cap yA$ for any two elements $x, y \in A$ with $AxA \cap AyA = 0$.

6) For any right denominator set S of the ring A, the right ring of fractions AS^{-1} is a PF ring.

7) The 2-generated right ideal $xA \oplus yA$ is a principal for any elements $x, y \in A$ with $AxA \cap AyA = 0$.

8) The 2-generated left ideal $Ax \oplus Ay$ is a principal for any two elements $x, y \in A$ with $(AxA) \cap (AyA) = 0$.

Proof. The equivalence of conditions 1), 2) and 3) follows from 3.2.3 and 1.3.1(b).

3) $\Rightarrow$ 5). Let $x, y \in A$ and $AxA \cap AyA = 0$. By 3), there exist $a, b \in A$ such that

$$a + b = 1\,, \quad xa = 0\,, \quad yb = 0\,, \quad (x + y)A = (x + y)bA + (x + y)aA = xbA \oplus yaA\,.$$

Therefore, $(x + y)A = (x + y)A \cap xA \oplus (x + y)A \cap yA$.

5) $\Rightarrow$ 4). Let $h_x \colon xA \oplus yA \to xA$ and $h_x \colon xA \oplus yA \to xA$ be natural projections. By assumption, $(x+y)A = h_x((x+y)A) \oplus h_y((x+y)A)$. Therefore, $x = h_x(x+y) \in (x+y)A$, $y = h_y(x + y) \in V$ and $(x + y)A = xA \oplus yA$.

The implication 4) $\Rightarrow$ 7) is obvious.

7) $\Rightarrow$ 3). By assumption, there exist $s, t, u, v \in A$ such that $x = (xs + yt)u$ and $y = (xs + yt)v$. Since $AxA \cap AyA = 0$, we have that $x = xsu$ and $0 = xsv = y(1 - tv) = yb$, where $b \equiv 1 - tv$. We set $a = 1 - b = tv \in A$. Since $v \in r(xs)$, it follows from 1.3.1(b) that $utv \in r(xs)$ and $xa = xsua = xsutv = 0$. We proved that there are $a, b \in A$ such that $a + b = 1$, $xa = 0$ and $yb = 0$.

The implication 6) $\Rightarrow$ 1) follows from the property that A is the right ring of fractions of A with respect to the set $S = \{1\}$.

1) $\Rightarrow$ 6). Let $f \colon A \to AS^{-1}$ be the canonical ring homomorphism with kernel $K = \{a \in \mid as = 0$ for some $s \in S\}$ and $h \colon A \to A/K$ be the natural epimorphism. By 1.3.6(a), $h(A)$ is a reduced ring and $h(A)h(S)^{-1} = AS^{-1}$ is the right ring of fractions of the ring $h(A)$ with respect to the right Ore set $h(S)$.

According to 3.2.3(e), it is sufficient to prove that $h(A)$ is a PF ring. Let $h(x)h(y) = h(0)$, where $x, y \in A$. Then $xys = 0$ for some element $s \in S$. Since A is a PF ring, there exist elements $a, b \in A$ such that $a + b = 1$, $xa = 0$ and $bys = 0$. Then $h(a) + h(b) = h(1)$, $h(x)h(a) = h(0)$, $by \in K$. Then $h(b)h(y) = h(0)$ and $h(A)$ is a PF ring.

1) $\Leftrightarrow$ 8). Since the equivalence of 1) $\Leftrightarrow$ 7) has been proven, the equivalence of 1) $\Leftrightarrow$ 8) follows from the property that condition 1) is left-right symmetrical. □

3.2.6. Let A be a reduced ring.

a. If for any elements $x, y \in A$ with $(AxA) \cap (AyA) = 0$ the right $A/r(xA \oplus yA)$-module $xA \oplus yA$ is flat, then A is a PF ring.

b. If any 2-generated right ideal X of the ring A is a flat $A/r(X)$-module, then A is a PF ring.

c. If the ring A is right distributive, then A is a PF ring.

d. If A is a right Bezout ring, then all submodules of flat right or left A-modules are flat.

e. Every finitely generated right ideal B of the ring A is a quasiprojective flat right A-module and a free cyclic right $A/r(B)$-module.

Proof. **a.** The assertion follows from 3.2.4.

b. Let $B = r(xA \oplus yA)$ and let $h\colon A \to A/B$ be the natural ring epimorphism. By 1.3.1(f), $h(A)$ is a reduced ring. By 1.3.1(d), $(AxA \oplus AyA) \cap \operatorname{Ker} h = 0$. In addition, $xA \oplus yA$ is a flat $h(A)$-module. Therefore, $h(AxA) \cap h(AyA) = h(0)$ and $h(xA) \cap h(yA)$ is a flat $h(A)$-module. Then the principal right ideal $h(xA)$ of the reduced ring $h(A)$ is a flat $h(A)$-module. By 3.2.4, $h(xA \oplus yA) = h((x+y)A)$. Then $xA \oplus yA \oplus B = (x+y)A \oplus B$. Then $xA \oplus yA = (x+y)A$, since $B \cap (xA \oplus yA) = 0$ and $(x+y)A \subseteq xA \oplus yA$. By 3.2.5, A is a PF ring.

b. The assertion follows from **a.**

c. By 3.2.5, A is a PF ring; see condition 6 of 3.2.5.

d. By 3.2.2, it is sufficient to prove that an arbitrary finitely generated right ideal B is a flat module. Since A is a right Bezout ring, B is a principal right ideal. By 3.2.5, A is a PF ring. By 3.2.3(b), B is a flat module.

e. Since A is a right Bezout ring, B is a principal right ideal of the ring A. By 1.3.1(h), B_A is a quasiprojective A-module and free cyclic right $A/r(B)$-module. By **d**, B_A is a flat A-module. □

3.2.7 Properties of PF rings. For a ring A, the following conditions are equivalent.

1) A is a PF ring.

2) All submodules of any H-torsion-free right A-module are H-torsion-free.

2') All submodules of any H-torsion-free left A-modules are H-torsion-free.

3) In A, all principal right ideals are H-torsion-free.

3') In A, all principal left ideals are H-torsion-free.

Proof. It is sufficient to prove the equivalence of conditions 1), 2) and 3).

1) $\Rightarrow$ 2). Let M_A be an H-torsion-free module, N be a submodule in M, and let B be a principal left ideal of A. By 3.2.3(b), ${}_AB$ is a flat module, Therefore, a natural group homomorphism $f\colon N \otimes B \to M \otimes B$ is a monomorphism. Since M is an H-torsion-free module, it follows from 3.1.4 that a natural group homomorphism $g\colon M \otimes B \to MB$ is an isomorphism. Let $h\colon N \otimes B \to NB$ be a natural group epimorphism. Since $h = gf$, we have that h is an isomorphism. By 3.1.4, N is an H-torsion-free module.

2) $\Rightarrow$ 3). By 3.1.6, A_A is an H-torsion-free module.

3) $\Rightarrow$ 1). Let $x, y \in A$ and $xy = 0$. Since xA is an H-torsion-free module, it follows from 3.1.4 that there exist elements $x_1, \dots, x_k \in xA$ and $a_1, \dots, a_k \in A$ such that

$x = \sum_{i=1}^{k} x_i a_i$ and $a_i y = 0$ for any i. There exist elements $y_1, \dots, y_k \in A$ such that $x_i = xy_i$ for $i = 1, \dots, k$. We set $a = 1 - \sum_{i=1}^{k} y_i a_i \in A$ and $b = \sum_{i=1}^{k} y_i a_i \in A$. Then $by = 0$,

$$xa = x - xb = x - x\sum_{i=1}^{k} y_i a_i = x - \sum_{i=1}^{k} x_i a_i = 0\,. \qquad \square$$

3.2.8 Two-generated flat right ideals, I. Let A be a ring.

a. Let $a_1, a_2, b_1, b_2 \in A$, $a_1 b_1 = a_2 b_2$ and let $a_1 A + a_2 A$ be a flat A-module. Then there exist elements $f_{ij} \in A$, $1 \le i, j \le 2$, such that

$$a_1 f_{11} = a_2 f_{21}\,, \quad (1 - f_{11}) b_1 = f_{12} b_2\,, \quad a_1 f_{12} = a_2 f_{22}\,, \quad (1 - f_{22}) b_2 = f_{21} b_1\,.$$

b. In A, all 2-generated right (left) ideals are flat if and only if for any $a_1, a_2, b_1, b_2 \in A$ with $a_1 b_1 = a_2 b_2$, there exist elements $f_{11}, f_{12}, f_{21}, f_{22} \in A$ such that $a_1 = a_1 f_{11} + a_2 f_{21}$, $a_2 = a_1 f_{12} + a_2 f_{22}$, $f_{11} b_1 = f_{12} b_2$ and $f_{21} b_1 = f_{22} b_2$.

Proof. **a.** Let X_A be a free A-module of rank 2 with basis $\{x_1, x_2\}$, $y \equiv x_1 b_1 + x_2 b_2$, $B \equiv a_1 A + a_2 A$ and $g\colon X_A \to B_A$ is an epimorphism such that $g(x_1 c_1 + x_2 c_2) = a_1 c_1 - a_2 c_2$ for any $c_1, c_2 \in A$. We set $Y = \operatorname{Ker} g$. Since $y \in Y$ and B is a flat module, it follows from 3.1.9 that $f(y) = y$ for some homomorphism $f\colon X \to Y$. Let $f_{ij} \in A$, $1 \le i, j \le 2$ and

$$f(x_1) = x_1 f_{11} + x_2 f_{21} \in Y\,, \quad f(x_2) = x_1 f_{12} + x_2 f_{22} \in Y\,. \quad \text{Then}$$
$$0 = g(f(x_1)) = g(f(x_2))\,, \quad a_1 f_{11} = a_2 f_{21}, a_1 f_{12} = a_2 f_{22}\,,$$
$$y = f(y) = f(x_1 b_1 + x_2 b_2) = x_1(f_{11} b_1 + f_{12} b_2) + x_2(f_{21} b_1 + f_{22} b_2)\,,$$
$$b_1 = f_{11} b_1 + f_{12} b_2\,, \quad b_2 = f_{21} b_1 + f_{22} b_2\,,$$
$$(1 - f_{11}) b_1 = f_{12} b_2\,, \quad (1 - f_{22}) b_2 = f_{21} b_1\,.$$

b. The proof is similar to the proof of **a**, as it uses 3.1.9. $\square$

3.2.9 Two-generated flat right ideals, II. Let A be a ring such that all 2-generated right ideals of A are flat.

a. For any elements $u, v, w, z \in A$ with $uv = zw$, there exist elements $f, g, h \in A$ such that $uf = zg$ and $(1 - f)v = hw$.

b. For any elements $u, v, w \in A$ with $uv = vw$, there exist elements $f, g, h \in A$ such that $uf = vg$ and $(1 - f)v = hw$.

c. For any elements $u, v, z \in A$ with $uv = zu$, there exist elements $f, g, h \in A$ such that $uf = zg$ and $(1 - f)v = hu$.

d. If A is a local ring, then A is a domain. In addition, if B, C are two principal right ideals in A and $B \cap C \ne 0$, then either $B \subseteq C$ or $C \subseteq B$.

e. If A is a right uniform local ring, then A is a right uniserial domain.

Proof. **a.** The assertion follows from 3.2.7(a).

b, c. The assertions follow from **a.**

d. By 3.2.3(d), A is a domain. Let $B = uA$ and $C = zA$. We assume that $B \not\subseteq C$. We have to prove that $C \subset B$. By assumption, $uv = zw \neq 0$ for some $u, v \in A$. By **a**, there exist elements $f, g, h \in A$ such that $uf = zg$ and $(1-f)v = hw$. The element f is not invertible, since otherwise $u \in C$, $B \subseteq C$; this is a contradiction. Since the ring A is local, $1 - f \in U(A)$. Then $v = (1-f)^{-1}hw$ and $zw = uv = u(1-f)^{-1}hw$. Since A is a domain, $z = u(1-f)^{-1}h$, whence $C = zA \subseteq uA = B$.

e. The assertion follows from **d.** □

3.2.10 PP rings. We recall that a ring A is called a *right PP ring* if for any element $x \in A$, there exists an idempotent $e \in A$ with $r(x) = eA$, i.e., the module xA is projective.

a. If A is a right or left PP ring, then A is a PF ring. There exist commutative PF rings that are not PP rings.

b. If A is a right Bezout ring which is a right PP ring, then the intersection of any two principal right ideals of the ring A is a principal right ideal.

Proof. **a.** For example, let A be a right PP ring and $xy = 0$, where $x, y \in A$. There exists an idempotent $e \in A$ with $r(x) = eA$. Since $y \in r(x)$, we have $ey = y$. Then $x = xe + x(1-e) = x(1-e)$. Then $a = 1 - e$ and $b = e$ are the required elements from the definition of a PF ring. We can verify that the ring from 2.2.3(c) is a commutative PF ring and is not a PP ring; see [70].

b. Let X, Y be two principal right ideals of the ring A. There exists an A-module epimorphism $h\colon X \oplus Y \to X + Y$ whose kernel is isomorphic to $X \cap Y$. Since A is a right Bezout ring, $X + Y$ is a principal right ideal. By assumption, $X + Y$ is a projective A-module. Therefore, the A-module $X \cap Y$ is isomorphic to a direct summand of the 2-generated module $X \oplus Y$. Consequently, $X \cap Y$ is a 2-generated right ideal of the right Bezout ring A. Therefore, $X \cap Y$ is a principal right ideal. □

3.2.11 Reduced PP rings. For a ring A, the following conditions are equivalent.

1) A is a right or left PP ring without noncentral idempotents.

2) A is a right and left PP reduced ring.

3) A is a reduced ring and for any its finitely generated ideal B, there exists a central idempotent $e \in A$ such that $r(B) = \ell(B) = eA = Ae$.

4) In the ring A, every element is the product of a central idempotent and a nonzero divisor.

Proof. The implication 4) $\Rightarrow$ 2) is directly verified.

The implication 2) $\Rightarrow$ 1) follows from 1.3.1(e).

1) ⇒ 4). Let A be a right PP ring without noncentral idempotents, $x \in A$ and $x^2 = 0$. Then $r(x) = eA$ for some central idempotent $e \in A$ and $x \in r(x) = eA$. Therefore, $x = ex = xe = 0$ and A is a reduced ring. By 1.3.1(d), $\ell(x) = r(x) = eA = Ae$. Therefore, A is a left PP ring.

Let $a \in A$. Then $r(a) = \ell(a) = eA = Ae$ for some central idempotent $e \in A$. We set $f = 1 - e$ and $d = (1 - e)a + e$. Then f is a central idempotent and $a = fd$. Let $b \in \ell(d)$. Then $0 = db = (1 - e)ab + eb$ and

$$eb = -(1-e)ab = -ab \in eA \cap (1-e)A = 0\,, \quad eb = 0\,, \quad b \in r(a) = eA\,,$$
$$b = eb = 0\,, \quad \ell(d) = 0\,.$$

By 1.3.1(d), $r(d) = \ell(d) = 0$.

2) ⇒ 3). Let $B = \sum_{i=1}^{n} Ab_iA$ and let e_i be central idempotents of A such that $r(b_i) = e_iA$ for $i = 1, \ldots, n$. We set $e = e_1 \cdot \cdots \cdot e_n$. By 1.3.1(d), $r(B) = \ell(B) = \cap(e_iA) = eA$.

3) ⇒ 1). By 1.3.1(e), A is a ring without noncentral idempotents. Let $x \in A$. By assumption, there exists a central idempotent $e \in A$ such that $r(AxA) = eA$. Since $r(x) = r(AxA)$ by 1.3.1(d), we have that A is a right PP ring. □

3.2.12 Arithmetical rings and flat modules. If A is an invariant arithmetical ring and M is a right A-module, then M is flat if and only if M is a Hattori torsion-free module.

Proof. If M is a flat module, then M is a Hattori torsion-free module by 3.1.6.

Let M be a Hattori torsion-free module. By 1.1.7(c), $MB \cap MC = M(B \cap C)$ for any two ideals B, C of the ring A. By 3.1.6, M is a flat module. □

3.3 Modules close to projective

3.3.1 Quasiprojective modules. A module M is said to be *quasiprojective* if M is projective with respect to itself, i.e., for any epimorphism $h: M \to \overline{M}$ and each homomorphism $\overline{f}: M \to \overline{M}$, there exists a homomorphism $f: M \to M$ with $\overline{f} = hf$.

a. It is directly verified that all projective or semisimple modules are quasiprojective. On the other hand, if a positive integer n is divided by the square of a prime integer, then the finite cyclic group $\mathbb{Z}/n\mathbb{Z}$ is a quasiprojective nonprojective nonsemisimple module over the ring of integers $\mathbb{Z}$.

b. In [171, Proposition 7.7], it is proved that all finitely generated submodules of a distributive module over a commutative ring are quasiprojective. This implies the following later result from [2]: All finitely generated ideals of a commutative arithmetical ring A are quasiprojective A-modules. The last result also follows from Theorem 3A.

The following example shows that the converse assertion is not true.

c. There exists a finite commutative nonarithmetical ring A such that all ideals are quasiprojective.
Indeed, let F be a finite field, A be the factor ring of the polynomial ring $F[x, y]$ module the ideal generated by x^2 and y^2, and let $h\colon F[x, y] \to A$ be the natural epimorphism. Since all proper ideals of the ring A are semisimple and the modules A_A and 0_A are projective, all ideals of the ring A are quasiprojective. In addition, A is a finite local ring that is not arithmetical, since

$$\begin{aligned}(\overline{x}+\overline{y})A &= (\overline{x}+\overline{y})A \cap (\overline{x}A + \overline{y}A)\\ &\neq (\overline{x}+\overline{y})A \cap \overline{x}A + (\overline{x}+\overline{y})A \cap \overline{y}A = \overline{0}\,,\end{aligned}$$

where $\overline{x}, \overline{y}, \overline{0}$ are natural images of the polynomials $x, y, 0$ from the ring $F[x, y]$.

d. There exist finite noncommutative rings satisfying Theorem 3A.
Let F be an additive Abelian group that is the direct sum of two cyclic groups of order 2 with generators f and g, respectively. We define a multiplication on F such that $f^2 = g$, $g^2 = f$, $fg = gf = f + g$. Then F turns into a finite field consisting of four elements $0, f + g, f, g$ with identity element $f + g$. Let α be an automorphism of the additive group F such that $\alpha(f) = g$ and $\alpha(g) = f$. It is directly verified that α is a nonidentity automorphism of the field F. Let $F[x, \alpha]$ be the skew polynomial ring over F such that the F-coefficients are written to the left of the monomials x^n and $xf = \alpha(f)$ for all $f \in F$. We denote by A the factor ring of the ring $F[x, \alpha]$ modulo the ideal generated by x^2. It is directly verified that A is a finite invariant ring which contains the unique proper one-sided ideal. Therefore, A is a finite invariant noncommutative arithmetical ring A such that all ideals are quasiprojective.

3.3.2 Finitely generated quasiprojective modules. Let A be a ring and let X be a finitely generated quasiprojective right A-module.

a. If X contains a finite subset $\{x_1, \dots, x_n\}$ such that $r(\{x_1, \dots, x_n\}) = 0$, then X is a projective module.

b. If X contains a finite subset $\{x_1, \dots, x_n\}$ such that $r(\{x_1, \dots, x_n\}) = r(X)$, then X is a projective $A/r(X)$-module.

c. If the ring A is right invariant, then X is a projective $A/r(X)$-module.

Proof. **a.** The assertion is well known; e.g., see [166, 2.12(2)].

b. It is directly verified that X is a finitely generated quasiprojective right $A/r(X)$-module. Now the assertion **b** follows from assertion **a** applied to the $A/r(X)$-module X.

c. Let $X = \sum_{i=1}^{n} x_iA$. Since the ring A is right invariant, $r(x_i) = r(x_iA)$ for any i. Therefore,

$$r(X) = r(x_1A) \cap \dots \cap r(x_nA) = r(\{x_1, \dots, x_n\})\,.$$

By **b**, X is a projective $A/r(X)$-module. □

3.3.3 Invariant semiprime rings. Let A be a right invariant semiprime ring.

a. Every prime factor ring of the ring A is a right uniform domain and every semiprime factor ring of the ring A is a right invariant, right localizable reduced ring.

b. If all 2-generated right ideals of the ring A are flat, then for any its maximal right ideal M, the right localization A_M is a right uniserial ring and the ring A is right distributive. In particular, A is an arithmetical ring.

c. If all 2-generated right ideals of the ring A are quasiprojective, then the ring A is right distributive. In particular, A is an arithmetical ring.

Proof. The right invariant semiprime ring A is reduced. By 1.3.7(e) and 1.3.8(a), the ring A_M exists, it is a local ring and $J(A_M) = M_M$. Let P be an ideal of the ring A, which is the kernel of the canonical ring homomorphism $f\colon A \to A_M$, and let $h\colon A/P$ be the natural epimorphism.

a. The remaining assertions of **a** are directly verified, with the use of 1.3.1(c).

b. By 1.3.8(d), it is sufficient to prove that A_M is a right uniserial ring.

By 3.1.11(m), all 2-generated right ideals of the ring A_M are flat. By 3.2.3(d), A_M is a domain. Therefore, P is a completely prime ideal. By **a**, the right invariant domain A/P is right uniform. Since A_M is the right ring of fractions of the ring $h(A)$ with respect to its right Ore set $h(A)\setminus h(M)$, the local domain A_M is right uniform by 3.2.3(c). By 3.2.9(e), A_M is a right uniserial domain.

c. By 1.3.8(d), it is sufficient to prove that A_M is a right uniserial ring.

By 3.3.2(c), every 2-generated right ideal Y of the ring A is a projective $A/r(Y)$-module. By 3.2.6(b), A is a PF ring. By 3.2.5, A_M is a local PF ring. By 3.2.3(d), A_M is a domain. Therefore, P is a completely prime ideal. By **a**, the right invariant domain A/P is right uniform. Since A_M is the right ring of fractions of the ring $h(A)$ with respect to its right Ore set $h(A) \setminus h(M)$, the local domain A_M is right uniform by 3.2.3(c).

Let X_M be an arbitrary nonzero 2-generated right ideal of the ring A_M. There exists a nonzero 2-generated right ideal X of the ring A such that $f(X)A_M = X_M$. Since $X_M \neq 0$, we have $h(X) \neq h(0)$ and $h(X)h(r(X)) = h(0)$ and $h(A)$ is a domain. Therefore, $r(X) \subseteq P \subseteq M$. Let $\bar{h}\colon A \to A/r(X)$ be the natural ring epimorphism. The ring $\bar{h}(A)$ is right invariant. By 1.3.1, $\bar{h}(A)$ is a reduced ring. By **a**, $\bar{h}(A)$ is a right invariant, right localizable ring. Since $r(X) \subseteq \operatorname{Ker} f = P \subseteq M$, it is directly verified that the ring A_M is the right localization of ring $\bar{h}(A)$ by its maximal right ideal $\bar{h}(M)$. Let $\bar{f}\colon \bar{h}(A) \to A_M$ be the canonical ring homomorphism of this right localization consistent with the canonical homomorphism $f\colon A \to A_M$. By assumption, the module X_A is quasiprojective. By 3.3.2(c), the right $\bar{h}(A)$-module $\bar{h}(X)$ is projective, i.e., all 2-generated right ideals of the ring $\bar{h}(A)$ are projective. In addition, A_M is the right localization of the ring $\bar{h}(A)$ by its maximal right ideal $\bar{h}(M)$. By **b**, A_M is a right uniserial ring. □

3.3.4 Endomorphism-liftable and π-projective modules. A module M is said to be *endomorphism-liftable* or *skew-projective* if for any epimorphism $h\colon M \to \overline{M}$ and each homomorphism $\overline{f}\colon \overline{M} \to \overline{M}$, there exists an endomorphism f of the module M with $\overline{f}h = hf$.

A module M is said to be *π-projective* if for any epimorphism $h\colon M \to \bar{M}$ and each idempotent endomorphism $\bar{f}$ of the module $\bar{M}$, there exists an endomorphism f of the module M with $\bar{f}h = hf$.

a. Let Q be a module and N a fully invariant submodule in Q. If Q is an endomorphism-liftable (resp., quasiprojective) module, then it is directly verified that Q/N is an endomorphism-liftable (resp., quasiprojective) module.

b. For any prime integer p, the quasicyclic Abelian group $\mathbb{Z}(p^\infty)$ is an endomorphism-liftable nonquasiprojective module over the ring of integers $\mathbb{Z}$.

c. Every uniserial module is π-projective.
The ring $\mathbb{Z}$ is a π-projective nonuniserial $\mathbb{Z}$-module.

d. Every endomorphism-liftable module is π-projective. If A is an incomplete discrete valuation domain and Q is the field of fractions of the domain A, then Q is a π-projective A-module which is not endomorphism-liftable.

e. Every quasiprojective module M is an endomorphism-liftable module.

f. Let A be a right invariant ring, M be a distributive right A-module and let X be a finitely generated endomorphism-liftable submodule in M. Then X is a quasiprojective module.

g. Let M be a π-projective module and X, Y be two submodules in M with $X + Y = M$. Then there exist homomorphisms $f\colon M \to X$ and $g\colon M \to Y$ such that

$$f(Y) + g(X) \subseteq X \cap Y\,, \quad (f + g - 1_M)(M) \subseteq X \cap Y\,,$$
$$M = (f + g)(M) + X \cap Y\,, \quad \operatorname{Ker}(f + g) \subseteq X \cap Y\,.$$

Proof. The assertions **a, b, c,** and **d** are directly verified.

e. Let $h\colon M \to \overline{M}$ be an epimorphism and let $\overline{f}$ be an endomorphism of the module $\overline{M}$. Then $\overline{f}h$ is a homomorphism from M into $\overline{M}$. Since M is quasiprojective, there exists an endomorphism f of the module M such that $hf = \overline{f}h$. Therefore, M is an endomorphism-liftable module.

f. Let X/Y be a factor module of the module X, $h\colon X \to X/Y$ be the natural epimorphism, and let $\overline{f}\colon X \to X/Y$ be some homomorphism. By 1.1.6(b), there exists an ideal B of the ring A with $Y = XB$. Then $(X/Y)B = (X/XB)B = 0$,

$$\overline{f}(Y) = \overline{f}(XB) = \overline{f}(X)B \subseteq (X/Y)B = 0$$

and $Y \subseteq \operatorname{Ker}\overline{f}$. Therefore, the homomorphism $\overline{f}$ induces the homomorphism $f_1\colon X/Y \to X \to Y$ and $f_1h = \overline{f}$. Since X is an endomorphism-liftable module, there exists a homomorphism $f: X \to X$ such that $hf = f_1 h = \overline{f}$. Therefore, the module X is quasiprojective.

g. Let $h: M \to M/(X \cap Y)$ be the natural epimorphism. Since $h(M) = h(X) \oplus h(Y)$, there exist natural projections $\overline{f}: h(M) \to h(X)$ and $\overline{g}: h(M) \to h(Y)$. Since M is π-projective, $\overline{f}h = hf$ and $\overline{g}h = hg$ for some $f, g \in \operatorname{End} M$. Therefore,

$$f(M) \subseteq X, \quad g(M) \subseteq Y, \quad f(Y) + g(X) \subseteq X \cap Y.$$

Since $(\overline{f} + \overline{g} - 1_{h(M)}) = 0$, we have $(f + g - 1_M)(M) \subseteq X \cap Y$. Since $M = (f + g)(M) + (f + g - 1_M)(M)$, we have $M = (f+g)(M) + X \cap Y$. If $x \in \operatorname{Ker}(f+g)$, then $x = (1_M - f - g)(x) \in X \cap Y$. Then $\operatorname{Ker}(f + g) \subseteq X \cap Y$. □

3.3.5 Remark. For a ring A, the following conditions are equivalent.

1) $aAb = 0$ for any elements $a, b \in A$ such that $ab = 0$.

2) For any element $a \in A$, the right ideal $r(a)$ is an ideal in A.

3) For any element $a \in A$, the right ideal $r(a)$ is an ideal in A if and only if for any element $b \in A$< the left ideal $\ell(b)$ is an ideal in A.

In 3.3.5, the equivalence of 1) ⇔ 2 is directly verified. Since condition 1) is left-right symmetrical, conditions 1) and 3) are also equivalent.

3.3.6 Remark. Let A be a noncommutative ring and let $R = A[x]$ be the polynomial ring. If $a, b \in A$ and $ab \neq ba$, then the right ideal $(a + x)R$ of the ring R is not a left ideal. In particular, R is not a left invariant ring.

Indeed, let's assume the contrary. Then there exist a polynomial $f \in R$ and an element $c \in A$ such that $b(a + x) = (a + x)(c + xf)$. Since $b(a + x)$ is a polynomial of degree 1, $f = 0$. Therefore, $b(a + x) = (a + x)c$. By equating the coefficients of x, we obtain that $b = c$ and $b(a + x) = (a + x)b$. Therefore, $ba = ab$; this is a contradiction.

3.3.7 Finitely endomorphism-extendable modules. Let n be a positive integer. A module M is said to be *n-endomorphism-extendable* if every endomorphism of any n-generated submodule in M can be extended to an endomorphism of the module M.
A module M is said to be *finitely endomorphism-extendable* if every endomorphism of any finitely generated submodule in M can be extended to an endomorphism of the module M.

a. Let A be a ring, M be a 2-endomorphism-extendable A-module, F be a fully invariant submodule in M, and let G_1, G_2 be two submodules in M with $G_1 \bigcap G_2 = 0$. Then $F \bigcap (G_1 + G_2) = F \bigcap G_1 + F \bigcap G_2$.

b. Let A be a 2-right endomorphism-extendable ring, F be an ideal of the ring A, and G_1, G_2 be two right ideals of the ring A with $G_1 \bigcap G_2 = 0$. Then $F \bigcap (G_1 + G_2) = F \bigcap G_1 + F \bigcap G_2$.

c. If A is a ring and every its factor ring is 2-right endomorphism-extendable, then A is an arithmetical ring.

d. Let M be an invariant finitely endomorphism-extendable module such that for every 2-generated submodule X in M, any automorphism of an arbitrary semisimple factor module $\overline{X}$ of the module X can be lifted to an endomorphism of the module X. Then the module M is distributive.

e. Let A be an 1-right endomorphism-extendable ring and $a \in A$. The right ideal $r(a)$ is an ideal of the ring A if and only if the left ideal Aa is an ideal. In particular, if a is an arbitrary left nonzero divisor in A, then Aa is an ideal.

f. The ring A is left invariant if and only if A is a right 1-endomorphism-extendable ring and $r(a)$ is an ideal in A for any element $a \in A$.

g. Let A be an 1-right endomorphism-extendable ring. Then A has the left classical ring of fractions Q and $aAa^{-1} \subseteq A$ for any nonzero divisor $a \in A$. If A has the right classical ring of fractions, then Q is the two-sided classical ring of fractions of the ring A. If A is a domain, then A is a left invariant domain which has the left classical division ring of fractions.

h. If a 1-right endomorphism-extendable ring A is a right invariant or reduced ring, then the ring A is left invariant.

i. Let A be a finitely right endomorphism-extendable ring. By **g**, A has the left classical ring of fractions Q and $sAs^{-1} \subseteq A$ for any nonzero divisor s in A. If $q \in Q$ and $q^{n+1} = q^n a_n + \cdots + q_1 a + a_0$ for some elements $a_0, \ldots, a_n \in A$, then $q \in s^{-1}As$ for some nonzero divisor s in A.

j. If A is a right 1-endomorphism extendable right Ore domain, then A is a left invariant, left Ore domain.

k. If A is a right 2-endomorphism extendable domain, then A is a left invariant, right and left Ore domain.

m. Let $\mathbb{H}$ be a noncommutative division ring of Hamiltonian quaternions and $A = \mathbb{H}[x]$. Then A is a right and left Ore domain, which is not right 1-endomorphism extendable.

Proof. **a.** Let $f = g_1 + g_2 \in F \bigcap (G_1 + G_2)$, where $g_1 \in G_1$ and $g_2 \in G_2$. We set $H = g_1 A \oplus g_2 A$. Let $h_i : H \to g_i A$ be natural projections. By assumption, endomorphisms h_i of the module H can be extended to endomorphisms $d_i \in \operatorname{End} M$ and $d_1(g_1) = d_2(g_2) = 0$, $d_1(g_2) = g_2$, $d_2(g_1) = g_1$. We have that F is a fully invariant submodule in M, $g_1 = d_2(g_1 + g_2) = d_2(f)$ and $g_2 = d_1(f)$. Therefore, $d_i(f) \in F$, $f = d_2(f) + d_1(f) \in F \cap G_1 + F \cap G_2$ and

$$F \cap (G_1 + G_2) \subseteq F \cap G_1 + F \cap G_2 \subseteq F \cap (G_1 + G_2)\,.$$

b. The assertion follows from **a**.

c. Let F, G_1, G_2 be ideals of the ring A, $F \subseteq G_1 + G_2$, and let $h\colon A \to A/(G_1 \cap G_2)$ be the natural ring epimorphism. By applying **b** to the ring $h(A)$, we obtain the relation $h(F)\bigcap(h(G_1) + h(G_2)) = h(F)\bigcap h(G_1) + h(F)\bigcap h(G_2)$. Therefore,

$$F \cap G_1 + F \cap G_2 + G_1 \cap G_2 = F + G_1 \cap G_2. \qquad (*)$$

By using $(*)$ and the modular law, we obtain that

$$F = F \cap (F \cap G_1 + F \cap G_2 + G_1 \cap G_2) = F \cap G_1 + F \cap G_2 + F \cap G_1 \cap G_2 = F \cap G_1 + F \cap G_2\,.$$

d. We assume that the module M is not distributive. By 1.1.5, there exists a 2-generated submodule X of the module M such that X has a factor module $X_1 = S_1 \oplus T_1$, where S_1, T_1 are isomorphic simple modules. Let $u\colon S_1 \to T_1$ be an isomorphism. We denote by f_1 an automorphism of the semisimple module X_1 such that $f_1(s + t) = u^{-1}(t) + u(s)$ for all $s \in S_1$ and $t \in T_1$. Let $h\colon X \to X_1$ be the natural epimorphism. By assumption, there exists an endomorphism f of the finitely generated module X such that $hf = \overline{f}h$. Let S, T be the complete pre-images in the module X modules S_1 and T_1, respectively. Since $f_1(S_1) = T_1 \not\subseteq S_1$, we have $f(S) \not\subseteq S$. Since M is a 2-finitely endomorphism-extendable module, the endomorphism f of the module X can be extended to some endomorphism g of the module M. Since M is an invariant module, $f(S) = g(S) \subseteq S$. This is a contradiction.

e. We assume that $r(a)$ is an ideal in A and $b \in A$. Then $r(a) \subseteq r(ab)$. Therefore, $f(a) = ab$ for some epimorphism $f\colon aA \to abA$. Then $f \in \mathrm{End}(aA)$. By assumption, $ta = f(a) = ab$ for some $t \in A$. Therefore, $Aa \supseteq aA$ and Aa is an ideal.

We assume that Aa is an ideal in A and $b \in A$. Then $ab = ca$ for some $c \in A$. Therefore, $ab\, r_A(a) = ca\, r_A(a) = 0$, $b\, r_A(a) \subseteq r_A(a)$ and $r(a)$ is an ideal.

f. Let A be a left invariant ring. It follows from 3.3.5 that $r(a)$ is an ideal in A for any element $a \in A$. Let $f \in \mathrm{End}\, aA$. Then $f(a) = ab$ for some element $b \in A$ and $f(ax) = f(a)x = abx$ for any $x \in A$. Since A is a left invariant ring, $ab = ca$ for some element $c \in A$. The relation $g(y) = cy$ defines an endomorphism $g \in \mathrm{End}\, A_A S$. In addition, $g(ax) = cax = abx = f(ax)$ for all $x \in A$. Therefore, g is an extension of the homomorphism f.

In **f**, the converse implication follows from **e**.

g. Let S be the set of all nonzero divisors in A, $s \in S$, $a \in A$. By **a**, $sa = bs$ for some $b \in A$. Therefore, S is a left Ore set. Therefore, A has the left classical ring of fractions. The inclusion $aAa^{-1} \subseteq A$ follows from **e.**

If A has the right classical ring of fractions, then Q is the two-sided classical ring of fractions of the ring A, since if an arbitrary ring has the right classical ring of fractions and the left classical ring of fractions, then these two rings of fractions can be naturally identified.

The last assertion from **g** follows from the first assertion.

h. By 3.3.5 and 1.3.1(d), all right annihilators of elements of the ring A are ideals. Therefore, **h** follows from **f.**

i. We denote by M the n-generated submodule $\sum_{i=1}^{n} q^i A$ of the module Q_A. Then $qM \subseteq M$. By the left-sided analog of 1.3.2(a), $sM \subseteq A$ for some nonzero divisor $s \in A$. The right ideal sM of A is finitely generated and $s \in sM$. In addition,

$$qs^{-1} \cdot sM = qM \subseteq M = s^{-1} \cdot sM \,.$$

Therefore, the rule $f(x) = sqs^{-1}x$ defines an endomorphism f of the finitely generated right ideal sM of the ring A. By assumption, f can be extended to an endomorphism g of the module A_A. We set $c = g(1)$. Then

$$sq = sqs^{-1} \cdot s = f(s) = cs \,, \quad q = s^{-1}cs \in s^{-1}As \,.$$

j. Let a, b be two nonzero elements of the domain A. The right Ore domain A has the right classical division ring of fractions Q, which contains the element aba^{-1}. The mapping $f\colon aA \to aA$, $f(ax) = (aba^{-1})(ax) = abx$, is an endomorphism of the principal right ideal aA. By assumption, there exists an element $c \in A$ such that $ca = f(a) = (aba^{-1})a = ab$. Therefore, A is a left invariant domain. Since A is a left invariant domain, A is a left Ore domain.

k. Let x be a nonzero element in A and let y be an element of the domain A with $xA \cap yA = 0$. There exists an endomorphism f of the 2-generated module $xA \oplus yA$ such that $f(x) = x$ and $f(y) = 0$. By assumption, there exists an element $a \in A$ such that $ax = f(x) = x$ and $ay = f(y) = 0$. Since $(a-1)x = 0$ and A is a domain, $a = 1$. Then $y = 0$. Therefore, A is a right Ore domain. Then A is a right quasi-continuous, right Ore domain. By **j**, A is a left invariant, left Ore domain.

m. It is well known that all right (left) ideals of the polynomial ring over a division ring are principal. Therefore, A is an Ore domain. Then the domain A is quasicontinuous. According to 3.3.6, A is not a left invariant ring. Now we use **k**. □

3.3.8. Let A be a right invariant ring and let M be an invariant, finitely endomorphism-extendable, endomorphism-liftable right A-module. The following conditions are equivalent.

1) All factor modules of the module M are finitely endomorphism-extendable modules.

2) All finitely generated submodules of the module M are endomorphism-liftable.

3) M is a distributive module and all its finitely generated submodules are quasiprojective.

Proof. The implication 3) $\Rightarrow$ 2) follows from the property that all quasiprojective modules are endomorphism-liftable.

2) $\Rightarrow$ 1). Let M/Y be a factor module of the module M, X/Y be a finitely generated submodule in M/X, and let $\overline{f}$ be an endomorphism of the module X/Y. There exists a natural isomorphism $u\colon X/Y \to X_1/(X_1 \cap Y)$, where X_1 is a finitely generated submodule in X and $\overline{f}$ naturally induces the endomorphism $\overline{f}_1$ of the finitely generated module

$X_1/(X_1 \cap Y)$ and $u^{-1}\overline{f}_1 u = \overline{f}$. Let $h\colon M \to M/(X_1 \cap Y)$ be the natural epimorphism and let $h_1 : X_1 \to X_1/(X_1 \cap Y)$ be the restriction of the epimorphism h to X_1. By assumption, the module $X_1/(X_1 \cap Y)$ is an endomorphism-liftable module. Therefore, there exists an endomorphism f_1 of the finitely generated submodule X_1 of M with $h_1 f_1 = \overline{f}_1 h_1$. Since M is a finitely endomorphism-extendable module, f_1 can be extended to an endomorphism g of the module M. Since M is an invariant module, $g(X) \subseteq X$, $g(Y) \subseteq Y$, $g(X_1) \subseteq X_1$, $g(X_1 \cap Y) \subseteq X_1 \cap Y$ and g induces the endomorphism f of the module M/Y. It is directly verified that f coincides with $\overline{f}$ on the module X/Y. Therefore, M/Y is a finitely endomorphism-extendable module.

1) $\Rightarrow$ 2). Let X be a finitely generated submodule of the module M, X/Y be a factor module of the module X, $\overline{f}$ be an endomorphism of the finitely generated module X/Y, $h\colon M \to M/Y$ be the natural epimorphism, and let $h_X\colon X \to X/Y$ be the restriction of the epimorphism h to the module X. By assumption, M/Y is a finitely endomorphism-extendable module. Therefore, $\overline{f}$ can be extended to an endomorphism $\overline{g}$ of the module M/Y. Since M is an endomorphism-liftable module, there exists an endomorphism g of the module M with $\overline{g}h = hg$. Since M is an invariant module, $g(X) \subseteq X$ and g induces the endomorphism f of the module X. Then

$$h_X f = (hg)_X = (\overline{g}h)_X = \overline{f} h_X \,.$$

Therefore, the module X is an endomorphism-liftable module.

2) $\Rightarrow$ 3). By 3.3.7(d), the module M is distributive. By 3.3.4(d), all finitely generated submodules of the module M are quasiprojective. □

3.3.9 Remark. Let A be a ring, A_1 and A_2 be two ideals of the ring A, and let x_1, x_2 be two elements of the ring A such that $x_1 - x_2 \in A_1 + A_2$. Then there exists an element $a \in A$ such that $a - x_1 \in A_1$ and $a - x_2 \in A_2$.
Indeed, let $x_1 - x_2 = a_1 - a_2$, where $a_1 \in A_1$, $a_2 \in A_2$. We denote by a the element $x_1 - a_1 = x_2 - a_2$. Then $a - x_1 = -a_1 \in A_1$, $a - x_2 = -a_2 \in A_2$. Therefore, a is the required element.

3.3.10 Arithmetical rings. For a ring A, the following conditions are equivalent.

1) A is an arithmetical ring.

2) For any ideals $A_1, \ldots, A_n$ of A and arbitrary elements $x_1, \ldots, x_n \in A$ such that $x_i - x_j \in A_i + A_j$ for all i, j, there exists an element $x \in A$ such that $x - x_i \in A_i$ for $i = 1, \ldots, n$.

3) For any ideals A_1, A_2, A_3 of the ring A and each element $d \in (A_1 + A_2) \cap (A_1 + A_3)$, there exists an element $x \in A_1$ with $x - d \in A_2 \bigcap A_3$.

Proof. 1) $\Rightarrow$ 2). For $n = 2$, the assertion follows from 3.3.9.

We assume that the assertion is true for $n - 1$. There exists an element $b \in A$ such that $b - x_i \in A_i$ for $i = 1, \ldots, n - 1$. In addition, $x_i - x_n \in A_i + A_n$ for $i = 1, \ldots, n - 1$

and $b - x_n = (b - x_i) + (x_i - x_n) \in \cap_{i=1}^{n-1}(A_i + A_n)$. Since the ring A is arithmetical, $b - x_n \in A_n + \bigcap_{i=1}^{n-1} A_i$, $b - x_n = a_n - d$, where $a_n \in A_n$, $d \in \bigcap_{i=1}^{n-1} A_i$. We set $a = b + d$. Then $a - x_n = an \in A_n$. In addition, $a - x_i = b - x_i + d \in A_i$ for $i = 1, \dots, n - 1$. Therefore, a is the required element.

2) ⇒ 3). The assertion follows from the property that 2) turns into 3) for $n = 3$, $x_1 = 0$, $x_2 = x_3 = d$.

3) ⇒,1). It is sufficient to prove that if M, B, C are ideals of the ring A and $d \in (M + B)\bigcap(M + C)$, then $d \in M + (B\bigcap C)$. Let $d = m_1 + b = m_2 + c$, where $m_1, m_2 \in M$, $b \in B$ and $c \in C$. By 3), there exists an element $x \in M$ such that $x - d \in B\bigcap C$. We set $y = d - x \in B \cap C$. Then $d = x + y \in M + B \cap C$. □

3.3.11 Theorem. For an invariant ring A, the following conditions are equivalent.

1) A is an arithmetical ring.

2) Every factor ring $\overline{A}$ of the ring A is a finitely endomorphism-extendable right $\overline{A}$-module and a finitely endomorphism-extendable left $\overline{A}$-module.

3) Every cyclic right A-module is a finitely endomorphism-extendable module.

4) A_A is a finitely endomorphism-extendable module and all its finitely generated submodules are endomorphism-liftable.

5) A_A is a finitely endomorphism-extendable distributive module and all its finitely generated submodules are quasiprojective.

Proof. We use the property that the free module A_A is an endomorphism-liftable module and the invariant ring A is an invariant right (left) A-module.

The equivalences 3) ⇔ 4) ⇔ 5) follow from 3.3.8.

The implications 2) ⇒ 3) and 5) ⇒ 1) are directly verified.

The implication 2) ⇒ 1) follows from 3.3.7(c).

1) ⇒ 2). Since the condition 1) is left-right symmetrical and is preserved by factor rings, it is sufficient to prove that ${}_AA$ is a finitely endomorphism-extendable left A-module. Let $M = \sum_{i=1}^{n} Am_i$ be an arbitrary finitely generated submodule in ${}_AA$ and $f \in \operatorname{End} {}_AM$. Since ${}_AM$ is a finitely generated distributive module over the invariant ring A, it follows from 1.1.6(c) that $f(Am_i) \subseteq Am_i$ for all i. Therefore, there exist elements $x_1, \dots, x_n \in A$ such that $f(m_i) = m_ix_i \in m_iA = Am_i$. Then $0 = (Am_i\bigcap Am_j)(x_i - x_j)$ for all i, j. Since $Am_i = m_iA$ and $Am_j = m_jA$, we have $x_i - x_j \equiv d_{ij} \in r(m_iA \cap m_jA)$. By 1.1.2, there exist elements $a_{ij}, b_{ij} \in A$ such that $1 = a_{ij} + b_{ij}$ and $m_iAa_{ij} + m_jAb_{ij} \subseteq m_iA\bigcap m_jA$. Then $a_{ij}d_{ij} \in r(m_iA) \equiv A_i$, $b_{ij} \in r(m_jA) \equiv A_j$ and $x_i - x_j = a_{ij}d_{ij} + b_{ij}d_{ij} \in A_i + A_j$. By 3.3.10, there exists an element $x \in A$ such that $x - x_i \in A_i$ for all i. Therefore, the relation $g(y) = yx$, $y \in A$, correctly defines an endomorphism g of the module ${}_AA$), which is an extension of the f. □

3.3.12 Theorem. An invariant semiprime ring A is an arithmetical ring if and only if every submodule of any flat A-module is a flat module.

Proof. If every submodule of any flat A-module is a flat module, then all 2-generated right ideals of the ring A are flat right A-modules, and A is an arithmetical ring by 3.3.3(b).

Let A be an invariant arithmetical semiprime ring. Then A is a distributive reduced ring. By 3.2.6(c), A is a PF ring. By 3.1.5(a), every principal right ideal is an H-torsion-free module. Therefore, an arbitrary right ideal M of the ring A has the following property: every cyclic submodule of the module M_A is an H-torsion-free module. By 3.1.4, M_A is a Hattori torsion-free module. By 3.2.13, M_A is a flat module. By 3.2.2, all submodules of any flat A-module are flat. □

3.3.13 Theorem. An invariant semiprime ring A is arithmetical if and only if every finitely generated ideal of the ring A is a quasiprojective right A-module.

Theorem 3.3.13 follows from 3.3.3(c) and Theorem 3.3.11.

3.3.14 The completion of the proof of Theorems 3A, 3B and 3C. Theorem 3A follows from of Theorems 3.3.12 and 3.3.13.

Theorem 3B follows from Theorem 3A and the property that by 3.2.3(f), we can assume that A is a reduced ring.

Theorem 3C follows from Theorem 3.3.11.

3.3.15. Let Q be an endomorphism-liftable module, f be an endomorphism of the module Q, and let N be a small submodule in Q such that Q/N is an endomorphism-liftable module. We set $X_n = N + \sum_{i=1}^{n} f^i(N)$, $X = \sum_{n=1}^{\infty} X_n = N + \sum_{i=1}^{\infty} f^i(N)$.
Then $X \subseteq J(Q)$ and there exists an endomorphism g of the module Q such that $(f - g)(Q) \subseteq X$, $(f - g)(X) \subseteq J(X)$ and $g(N) \subseteq N$.
For any positive integer n, we set $Y_n = \sum_{i=1}^{n} f^{i-1}(f - g)(N)$, $Y = \sum_{n=1}^{\infty} Y_n$.
Then $X_n = N + Y_n$, $X = N + Y$, Y_n is a small submodule in X_n, $Y \subseteq J(X)$ and $X = N + J(X)$.

Proof. Since $J(Q)$ is a fully invariant submodule in Q and $N \subseteq J(Q)$, we have $X = N + \sum_{i=1}^{n} f^i(N) \subseteq J(Q)$. Since $f(X) \subseteq X$, we have that f induces the endomorphism f' of the module Q/X. We denote by h a natural epimorphism $Q/N \to Q/X$. Since Q/N is an endomorphism-liftable module, there exists an endomorphism $\bar{g}$ of the module Q/N such that $h\bar{g} = f'h$.

We denote by t the natural epimorphism $Q \to Q/N$. Since Q is an endomorphism-liftable module, there exists an endomorphism g of the module Q such that $tg = \bar{g}t$. By construction, we have $ht(f - g)(Q) = 0$. Therefore, $(f - g)(Q) \subseteq X$. It is clear that $g(N) \subseteq N$. In addition, $X \subseteq J(Q)$. Therefore, $(f - g)(X) \subseteq J(X)$, since $(f - g)(Q) \subseteq X$ and the Jacobson radical passes to the Jacobson radical under module homomorphisms.

We use induction on n to prove that $X_n = N + Y_n$. Let $n = 1$. Since $g(N) \subseteq N$, we have

$$X_1 = N + f(N) + g(N) \supseteq N + (f - g)(N)$$
$$= N + Y_1 \supseteq N + g(N) + (f - g)(N) \supseteq X_1 .$$

Now we assume that $X_n = N + Y_n$. Then $X_{n+1} = X_n + f^{n+1}(N) =$

$$= X_n + f^n(f(N)) + f^n(g(N)) \supseteq N + Y_n + f^n(f - g)(N)$$
$$= N + Y_{n+1} \supseteq X_n + f^n g(N) + f^n(f - g)(N) \supseteq X_{n+1} .$$

Since $(f - g)(Q) \supseteq X$ and N is a small submodule in Q, we have that $(f - g)(N)$ is a small submodule in X. Therefore, $f^n(f - g)(N)$ is a small submodule in X for all n, since $f^n(X) \subseteq X$. Therefore, Y_n is a small submodule in X for all n and $Y \subseteq J(X)$. Since $X = N + Y$ and $Y \subseteq J(X)$, we have $X = N + J(X)$. □

3.3.16. Let's assume that the conditions of 3.3.15 hold and one of the following conditions hold.
1) Y is a small submodule in X.
2) $X = X_n$ for some n.
3) $J(X)$ is a small submodule in X.
4) The endomorphism f is either an idempotent endomorphism or a nilpotent endomorphism.
Then $f(N) \subseteq N$.

Proof. To prove the inclusion $f(N) \subseteq N$, it is sufficient to prove that each of the conditions 1)–4) implies the relation $X = N$.

1) If Y is a small submodule in X, then $X = N$, since $X = N + Y$.

2) If $X = X_n$, then $X = N + Y_n$, whence $X = N$, since Y_n is a small submodule in X by 3.3.15.

3) If condition 3) holds, then $X = N$ by 3.3.15.

4) If $f^2 = f$ or $f^n = 0$, then $X_2 = X$ or $X_n = X$, whence the condition 2) holds and $X = N$. □

3.3.17. Let M be an endomorphism-liftable module.

a. Q is a quasiprojective module, N is a small submodule in Q. If $M \cong Q/N$ and for every submodule X in $J(Q)$, the module $J(X)$ is a small submodule in X, then M is a quasiprojective module.

b. If M has a projective cover Q with kernel N and for every submodule X in $J(Q)$, the module $J(X)$ is a small submodule in X, then M is a quasiprojective module.

Proof. **a.** The condition 3) of 3.3.16 holds. By 3.3.16, N is a fully invariant submodule in Q. By 3.3.4(a), M is a quasiprojective module.

b. The assertion follows from **a.** □

3.3.18 Projective covers and perfect rings. Let Q be a projective module and N a small submodule of Q. If a module M is isomorphic to the module Q/N, then the module Q is called a *projective cover* of the module M with *kernel* N.

a. For a ring A, the following conditions are equivalent.
1) Every right A-module has projective cover.
2) $A/J(A)$ is an Artinian ring and $X \neq J(X)$ for every nonzero right A-module X.
3) $A/J(A)$ is an Artinian ring and $J(A)$ is a t-nilpotent left ideal.
Under the equivalent conditions 1)–3), the ring A is said to be *right perfect*.

b. If A is a right perfect ring, then $J(X)$ is a small submodule in X for every nonzero right A-module X.

The assertions from 3.3.18 are well known. For example, see **a** in [16, Theorem 28.4], and **b** follows from 3.3.17(b).

3.3.19 Theorem ([156]). Let A be a right perfect ring and M a right A-module. The following conditions are equivalent.

1) M is an endomorphism-liftable module.

2) M is a quasiprojective module.

3) M is a projective module over the factor ring $A/r(M)$.

Proof. The equivalence 2) $\Leftrightarrow$ 3) for right modules over right perfect rings is well known; see [16, Ex.16, p.203].

The implication 2) $\Rightarrow$ 1) is always true.

The implication 1) $\Rightarrow$ 2) follows from 3.3.17. □

3.3.20. In connection to 3.3.19, we remark that every quasicyclic Abelian group is an endomorphism-liftable nonquasiprojective $\mathbb{Z}$-module.

3.3.21 Open question. Let A be an invariant ring such that all 2-generated (all 1-generated) right ideals are flat. Is it true that A is a reduced ring?

4 Hermite rings and Pierce stalks

The main results of this section are Theorem 4A and 4B.

4A Theorem (Tuganbaev [183]). If A is a Bezout right PP ring without noncentral idempotents, then A is a Hermite ring.

4B Theorem (Tuganbaev [183]). If A is a Bezout ring such that every Pierce stalk is a serial ring, then A is a diagonalizable ring.

Remark. The completion of the proof of Theorems 4A and 4B is given in 4.2.9.

4.1 Hermite rings

4.1.1. Let A be a PF reduced ring that has the right classical ring of fractions Q.

a. A contains all idempotents of the ring Q.

b. If Q is a strongly regular ring, then A is a right and left PP ring.

c. If A is a right or left PP ring, then Q is a strongly regular ring.

Proof. **a.** Let $e = e^2 \in Q$. There exist two elements $a, b \in A$ such that b is a nonzero divisor in A and $e = xy^{-1}$. By 1.3.2(b), Q is a reduced ring. Since $e^2 = e$, we have $xy^{-1}x = x$ and $(xy^{-1} - 1)x = 0$. By 1.3.1(b), $x(xy^{-1} - 1) = 0$. Therefore,

$$x^2y^{-1} = x\,, \quad x^2 = xy\,, \quad x(y - x) = 0\,.$$

By 1.3.1(b), $(AxA) \cap (A(y - x)A) = 0$. By 3.2.5,

$$Ax \oplus A(y - x) = A(x + (y - x)) = Ay\,,$$
$$(Ax \oplus A(y - x))y^{-1} = A\,.$$

Therefore, $e = xy^{-1} \in A$.

b. Let $a \in A$. Since a is an element of the strongly regular ring Q, we have $a = eu = ue$, where e is a central idempotent of the ring Q and u is an invertible element of the ring Q. By **a**, $e \in A$ and $1 - e \in A$. It is clear that $r_Q(a) = (1 - e)Q$ and $r_A(a) = A \cap (1 - e)Q$. Since $1 - e \in A$, we have $r_A(a) = A \cap (1 - e)Q = (1 - e)A$. Therefore, $aA_A \cong A_A/r_A(a) \cong eA_A$, the module aA_A is projective, and A is a right PP reduced ring. By 3.2.11, A is a PP ring.

c. Let $q = ab^{-1} \in Q$, $a, b \in A$, and let b be a nonzero divisor in A. By 3.2.11, $a = ed$, where e is a central idempotent of the ring A and d is a nonzero divisor of the ring A. Then e is a central idempotent of the ring Q, the element db^{-1} is invertible in the ring Q and $q = edb^{-1}$. Since every element of the ring Q is the product of a central idempotent by an invertible element, Q is a strongly regular ring. □

https://doi.org/10.1515/9783110659825-004

4.1.2 Stably finite rings. A ring A is said to be *stably finite* if for any positive integer n, every right invertible matrix from the ring of matrices A_n is left invertible in A_n.

a. A right Noetherian ring is a stably finite ring.

b. Every direct product of stably finite rings is a stably finite ring.

c. Every subring of stably finite rings is a stably finite ring.

d. Every subring of any direct product of right Noetherian rings is stably finite. In particular, every subring of any direct product of division rings is stably finite.

e. If Q is a strongly regular ring and A is a subring in Q, then A is stably finite.

Proof. **a, b, c.** These assertions are well known; e.g., see [113, Proposition 1.9, Corollary 1.10, Proposition 1.13].

d. The assertion follows from **a, b** and **c.**

e. Since any strongly regular ring is a subdirect product of division rings, the assertion follows from **d.** □

4.1.3. Let A be a right Bezout reduced PP ring.

a. If a, b are two nonzero divisors of the ring A, then $aA \cap bA$ is a nonzero principal right ideal of the ring A.

b. If a, b are two nonzero divisors of the ring A, then there exist nonzero divisors a_1 and b_1 of the ring A such that $ab_1 = ba_1$ and $aA \cap bA = ab_1A = ba_1A$.

c. The ring A has the strongly regular right classical ring of fractions.

d. A is a stably finite ring.

Proof. **a.** By 3.2.10(b), $aA \cap bA$ is a principal right ideal.

We assume that $aA \cap bA = 0$. Since A is a right Bezout ring, there exist elements $x, y \in A$ such that $aA + bA = (ax + by)A$. Therefore, there exist elements $u, v \in A$ such that $a = (ax + by)u$ and $b = (ax + by)v$. Then

$$a(1 - xu) = byu \in aA \cap bA = 0\,, \quad b(1 - yv) = axv \in aA \cap bA = 0\,.$$

Since a and b are nonzero divisors,

$$1 - xu = yu = 1 - yv = xv = 0\,.$$

By 1.3.1(e), all right or left invertible elements of the ring A are invertible. Therefore, it follows from the relations $1 - xu = 1 - yv = 0$ that x, y, u, v are invertible elements. This contradicts the relations $yu = xv = 0$.

b. By **a**, $aA \cap bA = d_1A$, where d_1 is a nonzero element of the ring A. There exist elements $a_1, b_1 \in A$ such that $d_1 = ab_1 = ba_1$. By 3.2.11, $d_1 = ed$, where e is a nonzero central idempotent of the ring A and d is a nonzero divisor of the ring A.

If $e = 1$, then $d_1 = d = ab_1 = ba_1$ is a nonzero divisor, a_1 and b_1 is a nonzero divisors, and all have been proved.

We assume that $e \neq 1$. Then $1 - e$ is a nonzero central idempotent and $(1 - e)A$ is a nonzero reduced right Bezout ring such that all its principal right ideals are projective. In addition, $(1 - e)a$ and $(1 - e)b$ are nonzero divisors of the ring $(1 - e)A$ and $(1 - e)aA \cap (1 - e)bA = (1 - e)(aA \cap bA) = (1 - e)edA = 0$. This contradicts the assertion **a** applied to the ring $(1 - e)A$.

c. By 4.1.1(c), it is sufficient to prove that A has the right classical ring of fractions Q. It is sufficient to prove that for any elements $x, b \in A$, where b is a nonzero divisor, there exist elements $x_1, b_1 \in A$ such that b_1 is a nonzero divisor and $xb_1 = bx_1$.

By 3.2.11, $x = ea$, where e is a central idempotent of the ring A and a is a nonzero divisor of the ring A. By **b**, there exist nonzero divisors a_1 and b_1 of the ring A such that $ab_1 = ba_1$. We set $x_1 = ea_1$ Then $xb_1 = bx_1$.

d. The assertion follows from **c** and 4.1.2. □

4.1.4 ([9]). Let A be a right Bezout ring and let the right ideal $r(a)$ be finitely generated for any left zero divisor $a \in A$.

a. If V is a free right A-module with basis $\{v_l, \dots, v_n\}$, then every finitely generated submodule W in V is generated by n elements $w_k = v_k r_{kk} + \dots + v_n r_{kn}$, $r_{kk}, \dots, r_{kn} \in A$, $k = 1, \dots, n$.

b. The ring A_n of all $n \times n$ of matrices over A is a right Bezout ring and every finitely generated right ideal of the ring A is generated by a lower triangular matrix.

c. If M is an arbitrary $n \times m$ matrix over the ring A, then there exists a lower triangular $n \times m$ matrix D and two matrices P and Q such that $DQ = M$ and $MP = D$.

Proof. **a.** We set $V^k = v_k A + \dots + v_n A$. Let $I^{(k)}$ be the right ideal of the ring A consisting of all elements $r_k \in A$ such that there exist elements $v \in V^k \cap W$ of the form $v = v_k r_k + \dots + v_n r_n$. It is clear that $I^{(k)}$ is a right ideal.

First, we remark that if W is generated by elements $u_1, \dots, u_m$, then $I^{(1)}$ is a right ideal generated by m coefficients of v_1 in the expression of the elements $\{u_j\}$. Since A is a right Bezout ring, $I^{(1)} = r_{11}A$ for some $r_{11} \in A$ and there exists an element $w_1 \in W$ such that $w_1 = v_1 r_{11} + \dots + v_n r_{ln}$. It follows from the definition of $I^{(1)}$ that for every element $v \in W$, there exists an element $x \in A$ such that $v - w_1 x \in V^{(2)}$.

Next, we assert that $V^{(2)} \cap W$ is a finitely generated module. Indeed, let $w \in V^{(2)} \cap W$. We take an element $x_i \in A$ such that $u_i - w_1 x_i \in V^{(2)}$. Since set $\{u_i\}$ generates W, we have

$$w = \sum u_i y_i = \sum (u_i - w_1 x_i) y_i = +w_1 \cdot \sum x_i y_i \,. \qquad (*)$$

Since $w \in V^2$, we have that w has the nonzero coefficient of v_1. Therefore, the coefficient of v_1 in the right part of $(*)$ is equal to $r_{11} \cdot \sum x_i y_i = 0$, i.e., $\sum x_i y_i \in r(r_{11})$. In addition, if $a \in r(r_{11})$, then $wa \in V^{(2)} \cap W$. Therefore, $V^{(2)} \cap W$ is generated by the finite set

$$\{u_1 - w_1x_l, \dots, u_m - w_1x_m, w_1m_l, \dots, w_1m_t\},$$

where $\{m_l, \dots, m_t\}$ is a finite set of generators of the right annihilator $r(r_{11})$; this set is finite by assumption.

The remaining part of the proof is easy verified by the induction on n.

b. Let I be a finitely generated right ideal of the ring A_n and let V be a free n-generated right A-module with basis $\{v_l, \dots, v_n\}$, where v_i is a column of height n such that the coordinate with number i is equal to $1 \in A$ and the remaining coordinates are equal to $0 \in A$. We denote by W the submodule in V_A generated by all columns of matrices from I. Since I is a finitely generated right ideal of the ring A_n, we have that W is a finitely generated A-module. By **a**, the module W_A is generated by columns

$$w_1 = \begin{pmatrix} r_{11} \\ \vdots \\ r_{n1} \end{pmatrix}, \dots w_k = \begin{pmatrix} 0 \\ \vdots r_{kk} \\ \vdots \\ r_{nk}, \end{pmatrix} \dots w_n = \begin{pmatrix} 0 \\ \vdots 0 \\ \vdots \\ r_{nn} \end{pmatrix}.$$

We denote by P the lower triangular matrix formed by all columns w_j.

We prove that $I = PA_n$. Let E_{ik} be the matrix in A_n which has 1 on the position ij and 0s on the remaining positions. Since $w_j \in W$, we have that w_j appears in the row with the subscript $k = k(j)$ of some matrices $P_j \in I$. Therefore, $P = \sum P_jE_{kj} \in I$.

For any matrix $Q \in I$, the column with number j of the matrix Q is a linear combination $\sum w_ir_{ij}$. Therefore, $QE_{jj} = P(r)_j$, where $(r)_j$ is a matrix that contains

$$\begin{pmatrix} r_{1j} \\ \vdots \\ r_{nj} \end{pmatrix}$$

as the column with number j and zeros on the remaining positions. Therefore, $Q = \sum QE_{jj} = P(r_{ij})$; the proof is completed.

c. The assertion follows from the proof of **b**. □

4.1.5 Theorem ([9]). A domain A is a right Hermite ring if and only if A is a right Bezout ring.

Proof. Every right Hermite ring is a right Bezout ring by 2.2.3(b).

Now let A be a right Bezout domain and $a, b \in A$. We have to prove that $(a\ b)P = (d\ 0)$ for some invertible matrix $P \in A_2$.

If $a = 0$, then we can take the invertible matrix $P = \left(\begin{smallmatrix} 0 & 1 \\ 1 & 0 \end{smallmatrix}\right)$.

Let's assume now that $a \neq 0$. Then $aA + bA = dA$ for some nonzero of the element $d \in aA + bA$. We apply 4.1.4(c) to the right Bezout domain A and the matrix $M = \left(\begin{smallmatrix} a & b \\ 0 & 1 \end{smallmatrix}\right)$.

By 4.1.4(c), there exist 2×2 matrices $P, Q \in A_2$ such that

$$MP = \begin{pmatrix} d & 0 \\ u & v \end{pmatrix}; \quad \begin{pmatrix} d & 0 \\ u & v \end{pmatrix} Q = M\,.$$

Then $(a\,b)P = (d\,0)$ and $MPQ = M$. It remains to prove that the matrix P is invertible. Since $M = \left(\begin{smallmatrix} a & b \\ 0 & 1 \end{smallmatrix}\right)$ and a is a nonzero divisor in A, we have that M is a nonzero divisor in the matrix ring A_2. Therefore, PQ is the identity matrix. By 4.1.3(d), A is a stably finite ring. Therefore, the matrix P is invertible. □

4.1.6 Rings with diagonalizable square matrices. a. See [120, p.18]. If A is the ring of all lower triangular 2×2 matrices over a field, then all square matrices over A are diagonalizable, but A is not a diagonalizable ring.

b. If A is a Hermite ring and every square matrix over A is diagonalizable, then A is a diagonalizable ring.

Proof. We have to prove that an arbitrary rectangular $m \times n$ matrix M over the ring A is diagonalizable. By assumption, we can assume that $m \neq n$.

We assume that $m > n$. Since A is a left Hermite ring, there exists an invertible $m \times m$ matrix U such that UM is a upper triangular matrix. The matrix UM is of the form $\left(\begin{smallmatrix} S \\ O \end{smallmatrix}\right)$, where S is a square $n \times n$ matrix and O is a nonzero $(m-n) \times n$ matrix. By assumption, there exist invertible $n \times n$ matrices P and Y such that $PSY = D$ is a diagonal $n \times n$ matrix. Let E be the identity $(m-n) \times (m-n)$ matrix. We denote by X the invertible $m \times m$ matrix $\left(\begin{smallmatrix} P & 0 \\ 0 & E \end{smallmatrix}\right) U$. Then

$$X \cdot M \cdot Y = \begin{pmatrix} P & 0 \\ 0 & E \end{pmatrix} \cdot \begin{pmatrix} S \\ O \end{pmatrix} \cdot Y = \begin{pmatrix} D \\ O \end{pmatrix}$$

is a diagonal matrix.

The case $m < n$ is similarly considered. □

4.2 Pierce stalks

4.2.1 Pierce stalks, I. Let A be a ring and let $S(A)$ be the set of all proper ideals of the ring A generated by some sets of central idempotents. (Since $0 \in S(A)$, we have $S(A) \neq \varnothing$.)

By the Zorn lemma, the set $S(A)$ contains maximal elements which are called *Pierce ideals* of ring A. These Pierce ideals P form the set $\mathcal{P}(A)$ and the factor rings A/P are called *Pierce stalks* of ring A. On Pierce stalks, e.g., see [40, 41] and [170, Sections 11, 32].

Let P be a proper ideal of the ring A generated by some set $\{e_i\}_{i \in I}$ of central idempotents of the ring A, and let $h \colon A \to A/P$ be the natural epimorphism.

a. If the ring A/P is not a Pierce stalk of the ring A, then there exists a central idempotent $e \in A$ such that $P + eA$ and $P + (1 - e)A$ are proper ideals of the ring A generated by central idempotents, the ideals $P + eA$ and $P + (1 - e)A$ properly contain ideal P,

$$A = (P + eA) + (P + (1 - e)A) , \quad P = (P + eA)(P + (1 - e)A)$$
$$= (P + (1 - e)A)(P + eA) = (P + eA) \cap (P + (1 - e)A) ,$$

and the ring A/P is isomorphic to the direct product of nonzero rings $h(eA)$ and $h((1 - e)A)$.

b. For any finite subset P' of the ideal P, there exists a central idempotent $e \in P$ such that P' is contained in the ring direct factor eA of the ring A.

c. For any idempotent $\bar{f} \in A/P$, there exists an idempotent $f \in A$ with $h(f) = \bar{f}$.

d. If A/P is a Pierce stalk of the ring A and e is a central idempotent of the ring A, then either $e \in P$ or $1 - e \in P$.

e. If A/P is an indecomposable ring, then A/P is a Pierce stalk of the ring A.

f. There exists at least one Pierce stalk A/E with $P \subseteq E$.

g. If d is an element in A with zero right (left) annihilator and A/P is a Pierce stalk, then the element $d + P$ of the ring A/P has the zero right (left) annihilator.

h. If A is a PF ring, then A/P is a PF ring.

i. Any finite set of orthogonal idempotents $\{\bar{f}\}_{i=1}^{n}$ of the ring A/P can be lifted to some set $\{f_i\}_{i=1}^{n}$ of orthogonal idempotents of the ring A.

j. Any countable set of orthogonal idempotents $\{\bar{f}\}_{i=1}^{\infty}$ of the ring A/P can be lifted to some countable set $\{f_i\}_{i=1}^{\infty}$ of orthogonal idempotents of the ring A.

k. If A does not contain infinite sets of noncentral orthogonal idempotents, then the ring A/P does not contain infinite sets of noncentral orthogonal idempotents.

m. If A does not contain infinite sets of noncentral orthogonal idempotents and A/P is a Pierce stalk of the ring A, then A/P does not contain infinite sets of orthogonal idempotents.

n. If A is a right Bezout ring, then every Pierce stalk P of the ring A is a right Bezout ring.

Proof. **a.** Since the ideal P is generated by central idempotents and A/P is not a Pierce stalk, there exists a central idempotent $e \in A \setminus P$ such that $P + eA$ is a proper ideal in A. Then $P + (1 - e)A$ is a proper ideal in A, since otherwise $e \in e(P + (1 - e)A) \subseteq P$. The ideal $P + (1 - e)A$ properly contains P, since otherwise $1 = (1 - e) + e \in P + eA$. The remaining assertions are directly verified.

b. It is sufficient to prove that for any element $p \in P$, there exists a central idempotent $e \in P$ such that p belongs to the ring direct factor eA of the ring A. There exists a finite subset $\{e_1, \dots, e_n\}$ of the set $\{e_i\}_{i \in I}$ of central idempotents, which are generators for the ideal P, such that $p \in \sum_{i=1}^{n} e_iA$. There exists a central idempotent $e \in A$ with $eA = \sum_{i=1}^{n} e_iA$. Therefore, $p \in eA$

c. Let $\overline{f} = h(a)$, where $a \in A$. Then $a - a^2 \in P$. By **b**, there exists a central idempotent $e \in P$ with $a - a^2 \in eA$. Then $a - a^2 = ea - ea^2$ and $a(1-e) - (a(1-e))^2 = 0$. Therefore, $a(1-e)$ is an idempotent of the ring A and $h(a(1-e)) = h(a) = \overline{f}$.

d. We assume that $e \notin P$. Since A/P is a Pierce stalk of the ring A and e is a central idempotent, $P + eA = A$. Then $1 = ea + p$ for some $a \in A$ and $p \in P$. Then $1 - e = (1-e)(ea+p) = (1-e)p \in P$.

e. The assertion follows from **a**.

f. Let $\mathcal{E}$ be the set of all proper ideals in A which contain P and are generated by a central idempotents in A. Since $P \in \mathcal{E}$, we have $\mathcal{E} \neq \varnothing$. By the Zorn lemma, it is sufficient to prove that if E is the union of an ascending chain of ideals $E_i \in \mathcal{E}$, $i \in I$, then $E \in \mathcal{E}$. Since all ideals E_i are generated by central idempotents, E is generated by central idempotents. It remains to prove that $E \neq A$. We assume the contrary. Then $1 \in E = \bigcup_{i \in I} E_i$. Therefore, 1 is contained in some proper ideal E_i in A. This is a contradiction.

g. Let $r_A(d) = 0$, $a \in A$ and $da \in P$. By **b**, there exists a central idempotent $e \in P$ such that $da \in eA$. Then $da(1-e) = 0$. Since $r(d) = 0$, we have $a(1-e) = 0$ and $a = ae \in P$. Therefore, $r_{A/P}(d + P) = 0$.

h. Let $a, b \in A$ and $h(a)h(b) = 0$. Then $ab \in P$. By **b**, there exists a central idempotent $e \in P$ such that $ab \in eA$. Then $a(1-e)b = 0$. Since A is a PF ring, it follows from 3.2.3 that there exist elements $x, y \in A$ such that

$$x + y = 1\,, \quad a(1-e)x = 0\,, \quad y(1-e)b = 0\,,$$
$$h(x) + h(y) = h(1)\,, \quad h(a)h(x) = 0\,, \quad h(y)h(b) = 0\,.$$

By 3.2.3, A/P is a PF ring.

i. There exist elements $a_1, \dots, a_n \in A$ such that $\overline{f}_i = h(a_i)$ for all i. Then $a_i - a_i^2 \in P$, for all i, and $a_ia_j \in P$ for $i \neq j$. By **b**, there exists a finite subset $J = \{e_1, \dots, e_n\}$ in I such that the ideal $\sum_{i \in J}^{n} e_iA$ contains all elements $a_i - a_i^2$ and $a_ia_j \in P$, where $i, j \in J$, $i \neq j$. There exists a central idempotent $e \in P$ with $eA = \sum_{i \in J} e_iA$. We set $f_i = a_i(1-e)$, $i = 1, \dots, n$. Then $h(f_i) = h(a_i)$ for all i. In addition, all elements $f_i - f_i^2$ and e_ie_j $(i \neq j)$ are contained in the ideal $fA \cap (1-f)A = 0$. Therefore, $f_1, \dots, f_n$ are the required orthogonal idempotents of the ring A.

j. There exists a countable set $\{a_i\}_{i=1}^{\infty}$ of elements of the ring A such that $\overline{e}_i = h(a_i)$ for all i. Let $n \in \mathbb{N}$. By **i**, there exists a set of orthogonal idempotents $\{e_i\}_{i=1}^{n}$ of the ring A such that $h(e_i) = \overline{e}_i$ for $i = 1, \dots, n$. Now it is sufficient to prove that there is an idempotent $e_{n+1} \in A$ such that the idempotents $e_1, \dots, e_{n+1}$ are orthogonal and $h(e_{n+1}) = \overline{e}_{n+1}$. The ideal P contains elements $a_{n+1} - a_{n+1}^2$, e_ia_{n+1}, $a_{n+1}e_i$ for $i = 1, \dots, n$. There exist central idempotents $f_1, \dots, f_m \in P$ such that the ideal $\sum_{k=1}^{m} f_kA$ contains elements $a_{n+1} - a_{n+1}^2$, e_ia_{n+1}, $a_{n+1}e_i$ for $i = 1, \dots, n$. There exists a central idempotent f of the ring A such that $fA = \sum_{k=1}^{m} f_k A$. We set $e_{n+1} = a_{n+1}(1-f)$. Then $h(e_{n+1}) = h(a_{n+1})$. In addition, elements $e_{n+1} - e_{n+1}^2$, $e_i e_{n+1}$, $e_{n+1}e_i$ $(i \neq j)$ are contained in the ideal $fA \cap (1-f)A = 0$ for $i = 1, \dots, n$. Therefore, $e_1, \dots, e_{n+1}$ are orthogonal idempotents of the ring A.

k. The assertion follows from **j**.

m. We assume that the Pierce stalk A/P contains a countable set $\{\overline{e}_i\}_{i=1}^{\infty}$ of nontrivial orthogonal idempotents. Let $h\colon A \to A/P$ be the natural epimorphism. By **j**, A contains a countable set of orthogonal idempotents $\{e_i\}_{i=1}^{\infty} \subseteq A \setminus P$ such that $h(e_i) = \overline{e}_i$ for all i. Since A does not contain infinite sets of noncentral orthogonal idempotents, there exists a central idempotent $e \in \{e_i\}_{i=1}^{\infty}$. Since A/P is a Pierce stalk, $A = P + eA$. Then $h(e) = h(1)$; this contradicts the nontriviality of the idempotent $h(e)$ of the ring $h(A)$.

n. The assertion follows from the property that every factor ring of a right Bezout ring is a right Bezout ring. □

4.2.2 Pierce stalks, II. Let A be a ring, M be a right A-module, and let $\{A/P_i\}_{i\in I}$ be the set of all Pierce stalks of the ring A.

a. If $0 \neq m \in M$, then $m \notin MP_i$ for some Pierce stalk A/P_i.

b. $\cap_{i\in I}(MP_i) = 0$; in particular, A is the subdirect product of their Pierce stalks.

c. If N is a submodule in M and $m \notin N$, then $m \notin N + MP_i$ for some Pierce stalk A/P_i.

d. If N is a submodule in M, $m \in M$ and $m \in N + MP$ for every Pierce stalk A/P, then $m \in N$.

e. For each nonzero right A-module M, there exists at least one Pierce stalk A/P_i with $M \neq MP_i$.

f. For each indecomposable right A-module X, there exists a Pierce stalk A/P such that $P \subseteq r(X)$.

g. If all idempotents of any Pierce stalk A/P of the ring A are central, then all idempotents of the ring A are central.

h. If any Pierce stalk of the ring A does not have nontrivial idempotents, then all idempotents of the ring A are central.

Proof. **a.** Let $\{B_j\}_{j\in J}$ be the set of all proper ideals B_j in A such that B_j is generated by central idempotents in A and $m \notin MB_j$. The set $\{B_j\}$ is nonempty, since it contains the zero ideal. In addition, it is directly verified that $\{B_j\}$ contains the union of any ascending chain of its elements. By the Zorn lemma, $\{B_j\}$ contains a maximal element P. Then $m \notin MP$. It is sufficient to prove that A/P is a Pierce stalk of the ring A. Let's assume the contrary. By 4.2.1(a), $P = QS = SQ$ for some proper ideals Q and S such that Q and S properly contain P, $A = Q + S$ and Q, S are generated by central idempotents. Since P is a maximal element in $\{B_j\}$, we have $m \in MQ \cap MS = (MQ \cap MS)(Q + S) \subseteq MSQ + MQS = MP$. This is a contradiction.

b. The assertion follows from **a**.

c. We apply **a** to the nonzero element $m + N \in M/N$. By **a**, $m + N \notin (M/N)P_i = (N + MP_i)/N$ for some Pierce stalk A/P_i. Therefore, $m \notin N + MP_i$.

d, e. The assertions follow from **c** and **b**, respectively.

f. Let P be the ideal of A generated by the set of all central idempotents in A which are contained in the ideal $r(X)$. It is sufficient to prove that $P + eA = A$ for any central idempotent $e \in A \setminus P$. By the definition of the ideal P, we have $e \in A \setminus r(X)$. Therefore, Xe is a nonzero submodule of the indecomposable module $X = Xe \oplus X(1 - e)$. Then $X(1 - e) = 0$ and $1 - e \in r(X)$. Therefore, $1 - e \in P$. Then $A = (1 - e)A + eA \subseteq P + eA$.

g. Let $e = e^2 \in A$, $a \in A$, A/P be a Pierce stalk, and let $h\colon A \to A/P$ be the natural epimorphism. We have to prove that $ea - ae = 0$ for any element $a \in A$. By **b**, A is the subdirect product of their Pierce stalks. Therefore, it is sufficient to show that $h(ea - ae) = 0$. This is the case by assumption.

h. The assertion follows from **g**. □

4.2.3 Pierce stalks, III. Let A be a right PP ring such that all its idempotents are central.

a. Every Pierce stalk A/P of the ring A is a domain.

b. If A is a right Bezout ring, then every Pierce stalk of the ring A is a right Hermite ring.

Proof. **a.** Let A/P be a Pierce stalk and let $h\colon A \to A/P$ be the natural epimorphism. We assume that $h(a)h(b) = h(0)$, where $a, b \in A$ and $h(a) \neq h(0)$, i.e., $a \notin P$ and $ab \in P$. By 4.2.1(b), there exists a central idempotent $e \in P$ such that ab is contained in the ring direct factor eA of the ring A. Then $ab(1 - e) = 0$. By assumption, there exists a central idempotent f of the ring A that $r_A(a) = fA$. Therefore, $b(1 - e) = b(1 - e)f$ By 4.2.1(d), either $h(f) = h(0)$ or $h(f) = h(1)$, i.e., either $1 - f \in P$ or $f \in P$.

We assume that $1 - f \in P$. Then $h(f) = h(1)$, $h(a) = h(af) = h(0)$. This is a contradiction, since $h(a) \neq h(0)$.

We assume that $f \in P$. Then $b = be + b(1 - e)f \in P$, $h(b) = h(0)$ and A/P is a domain.

b. Let A/P be a Pierce stalk of the ring A. By **a** and 4.2.1(d), A/P is a right Bezout domain. By 4.1.5, A/P is a right Hermite ring. □

4.2.4 The images of integers in rings. When we consider an integer z as an element of some factor ring $\overline{A}$ of the ring A, then we mean the element $\alpha(z)$ under a natural unitary ring homomorphism $\alpha\colon \mathbb{Z} \to \overline{A}$.

4.2.5. Let A be a ring, $a_1, \dots, a_m \in A$, and let $f_1, \dots, f_k$ be polynomials with integral coefficients in noncommuting variables $x_1, \dots, x_m, y_1, \dots, y_n$. For each factor ring A/P, we denote by h_P the natural ring epimorphism $A \to A/P$. The following conditions are equivalent.

1) There exist elements $b_1, \dots, b_n$ of the ring A such that

$$f_j(a_1, \dots, a_m, b_1, \dots, b_n) = 0\,, \quad j = 1, \dots, k\,.$$

2) For every factor ring A/P of the ring A, there exist elements $\overline{b}_1, \dots, \overline{b}_n$ of the ring A/P such that A/P satisfies the relations

$$f_j(h_P(a_1), \dots, h_P(a_m), \overline{b}_1, \dots, \overline{b}_n) = h_P(0)\,, \quad j = 1, \dots, k\,.$$

3) For every Pierce stalk A/P of the ring A, there exist elements $\overline{b}_1, \dots, \overline{b}_n$ of the ring A/P such that A/P satisfies the relations

$$f_j(h_P(a_1), \dots, h_P(a_m), \overline{b}_1, \dots, \overline{b}_n) = h_P(0)\,, \quad j = 1, \dots, k\,.$$

Proof. The implications 1) $\Rightarrow$ 2) and 2) $\Rightarrow$ 3) are obvious.

3) $\Rightarrow$ 1). In this proof, we call the factor ring A/P *special* if there exist elements $\overline{b}_1, \dots, \overline{b}_n$ of the ring A/P such that the relations $f_i(h_P(a_1), \dots, h_P(a_m), \overline{b}_1, \dots, \overline{b}_n) = 0$, $i = 1, \dots, k$, are true in A/P. We denote by $\mathcal{E}$ the set of all proper ideals E in A such that A/E is not special. We denote by $\mathcal{E}^*$ the subset in $\mathcal{E}$ formed by all elements from $\mathcal{E}$ which are the ideals of A generated by some sets of central idempotents. We remark that the sets $\mathcal{E}$ and $\mathcal{E}^*$ can be empty.

We assume that the condition 1) is not true and the condition 3) is true. Since $0 \in \mathcal{E}^*$, we have $\mathcal{E}^* \neq \varnothing$. It is directly verified that the union of any ascending chain of ideals from $\mathcal{E}^*$ is contained in $\mathcal{E}^*$. By the Zorn lemma, $\mathcal{E}^*$ contains a maximal element P. By 3), it is sufficient to prove that A/P is a Pierce stalk. We assume the contrary. Then there exists a central idempotent $e \in A$ such that $P + eA$ and $P + (1 - e)A$ are proper ideals in A, $P+eA$ and $P+(1-e)A$ properly contain P, and $A/P \cong (A/(P+eA)) \times (A/(P+(1-e)A))$. The ideals $P + eA$ and $P + (1 - e)A$ are generated by some sets of central idempotents. In addition, $P + eA$ and $P + (1 - e)A$ are not contained in $\mathcal{E}^*$. Therefore, $A/(P + eA)$ and $A/(P + (1 - e)A)$ are special rings. Since $A/P \cong (A/(P + eA)) \times (A/(P + (1 - e)A))$, it is directly verified that A/P is a special ring. This is a contradiction. □

4.2.6. Let A be a ring, $\{P_i\}_{i \in I}$ be the set of all Pierce ideals of the ring A, and let $h_i : A \to A/P_i$ be natural ring epimorphisms. Sometimes, we write $\overline{a}$ instead of $h_i(a)$.

a. Let $M = \begin{pmatrix} a_{11} & a_{12} \\ a_{21} & a_{22} \end{pmatrix}$ be a 2×2 matrix, where $a_{11}, a_{12}, a_{21}, a_{22} \in A$, and for any $i \in I$, the matrix $M_i = \begin{pmatrix} h_i(a_{11}) & h_i(a_{12}) \\ h_i(a_{21}) & h_i(a_{22}) \end{pmatrix}$ is left invertible in the ring of all 2×2 matrices over the Pierce stalk A/P_i. Then the matrix M is left invertible in the ring 2×2 of matrices over A.

b. Let a_1, a_2 be two elements of the ring A. If for every $i \in I$, there exist elements $\overline{b}_{11}, \overline{b}_{12}, \overline{b}_{21}, \overline{b}_{22} \in A/P_i$ such that the matrix $M_i = \begin{pmatrix} \overline{b}_{11} & \overline{b}_{12} \\ \overline{b}_{21} & \overline{b}_{22} \end{pmatrix}$ is right invertible in the ring of all 2×2 matrices over A/P_i and the product of $(h_i(a_1), h_i(a_2))M_i$ is a row of length 2 with zero second element, then there exist elements $b_{11}, b_{12}, b_{21}, b_{22} \in A$ such that the matrix $M = \begin{pmatrix} b_{11} & b_{12} \\ b_{21} & b_{22} \end{pmatrix}$ is right invertible in the ring of all 2×2 matrices over A and the product of $(a_1, a_2)M$ is a row of length 2 with zero second element.

c. If every Pierce stalk A/P_i of the ring A is a right Hermite ring, then A is a right Hermite ring.

d. If n is a positive integer and every $n \times n$ matrix over any Pierce stalk A/P_i of the ring A is diagonalizable, then every $n \times n$ matrix over the ring A is diagonalizable.

e. If every Pierce stalk A/P_i of the ring A is a Hermite ring such that every square matrix over A/P_i is diagonalizable, then A is a diagonalizable ring.

f. If every Pierce stalk A/P_i of the ring A is a semilocal right Bezout ring, then A is a right Hermite ring.

g. If every Pierce stalk A/P_i of the ring A is a serial Bezout ring, then A is a diagonalizable ring.

h. If A is a Bezout ring such that every Pierce stalk is a serial ring, then A is a diagonalizable ring.

i. If every Pierce stalk A/P_i of the ring A is a right Bezout domain, then A is a right Hermite ring.

j. If A is a PF reduced ring such that the right annihilator of each element of A is a finitely generated right ideal, then A is a PP ring and every Pierce stalk of the ring A is a domain.

Proof. **a.** We consider the following polynomials f_1, f_2, f_3, f_4 in noncommuting variables $x_{11}, x_{12}, x_{21}, x_{22}, y_{11}, y_{12}, y_{21}, y_{22}$:

$$f_1 = y_{11}x_{11} + y_{12}x_{21} - 1\,, \quad f_2 = y_{11}x_{12} + y_{12}x_{22}\,,$$
$$f_3 = y_{21}x_{11} + y_{22}x_{21}\,, \quad f_4 = y_{21}x_{12} + y_{22}x_{22} - 1\,.$$

In the 2×2 matrix ring over A, the matrix M is left invertible if and only if there exist elements $b_{11}, b_{12}, b_{21}, b_{22}$ of the ring A such that $f_j(a_{11}, a_{12}, a_{21}, a_{22}, b_{11}, b_{12}, b_{21}, b_{22}) = 0$, $j = 1, 2, 3, 4$.

For any $i \in I$, the matrix $M_i = \begin{pmatrix} h_i(a_{11}) & h_i(a_{12}) \\ h_i(a_{21}) & h_i(a_{22}) \end{pmatrix}$ is left invertible in the ring 2×2 of matrices over the Pierce stalk A/P_i. Therefore, there exist elements $\overline{b}_{11}, \overline{b}_{12}, \overline{b}_{21}, \overline{b}_{22}$ of the ring A/P_i such that the relations $f_j(h_i(a_{11}), h_i(a_{12}), a_{21}), h_i(a_{22}), \overline{b}_{11}, \overline{b}_{12}, \overline{b}_{21}, \overline{b}_{22}) = h_i(0)$, $j = 1, 2, 3, 4$, are true in the ring A/P_i.

By 4.2.5, there exist elements $b_{11}, b_{12}, b_{21}, b_{22}$ of the ring A such that $f_j(a_{11}, a_{12}, a_{21}, a_{22}, b_{11}, b_{12}, b_{21}, b_{22}) = 0$, $j = 1, 2, 3, 4$. Therefore, the matrix M is left invertible in the ring of all 2×2 matrices over A.

b. We consider the following polynomials $f_1, f_2, f_3, f_4, f_5, f_6$ in noncommuting variables $x_1, x_2, z, y_{11}, y_{12}, y_{21}, y_{22}, y'_{11}, y'_{12}, y'_{21}, y'_{22}$:

$$\begin{aligned} f_1 &= x_1y_{11} + x_2y_{21} - z\,, & f_2 &= x_1y_{12} + x_2y_{22}\,,\\ f_3 &= y_{11}y'_{11} + y_{12}y'_{21} - 1\,, & f_4 &= y_{11}y'_{12} + y_{12}y'_{22}\,,\\ f_5 &= y_{21}y'_{11} + y_{22}y'_{21}\,, & f_6 &= y_{21}y'_{12} + y_{22}y'_{22} - 1\,. \end{aligned}$$

We consider an arbitrary Pierce stalk A/P_i. By assumption, there exist elements $\overline{d}, \overline{b}_{11}, \overline{b}_{12}, \overline{b}_{21}, \overline{b}_{22}, \overline{b}_{11}{}', \overline{b}_{12}{}', \overline{b}_{21}{}', \overline{b}_{22}{}' \in A/P_i$ such that

$$f_j(\overline{a}_1, \overline{a}_2, \overline{d}, \overline{b}_{11}, \overline{b}_{12}, \overline{b}_{21}, \overline{b}_{22}, \overline{b}_{11}{}', \overline{b}_{12}{}', \overline{b}_{21}{}', \overline{b}_{22}{}') = h_i(0)\,, \quad j = 1, 2, 3, 4, 5, 6\,.$$

By 4.2.5, there exist elements

$$d, b_{11}, b_{12}, b_{21}, b_{22}, b'_{11}, b'_{12}, b'_{21}, b'_{22} \in A$$

such that $f_j(a_1, a_2, d, b_{11}, b_{12}, b_{21}, b_{22}, b'_{11}, b'_{12}, b'_{21}, b'_{22}) = 0$, $j = 1, 2, 3, 4, 5, 6$. We denote by M and M' the 2×2 matrices $\begin{pmatrix} b_{11} & b_{12} \\ b_{21} & b_{22} \end{pmatrix}$ and $\begin{pmatrix} b'_{11} & b'_{12} \\ b'_{21} & b'_{22} \end{pmatrix}$, respectively. Then $M{\cdot}M'$ is the identity $2{\times}2$ matrix. Therefore, the matrix M is right invertible. In addition, $(a_1, a_2)M = (d, 0)$ is a row of length 2 with zero as the second element.

c. Let a_1, a_2 be two elements of the ring A. Since every Pierce stalk A/P_i of the ring A is a right Hermite ring, we have that for every $i \in I$, there exist elements $\overline{b}_{11}, \overline{b}_{12}, \overline{b}_{21}, \overline{b}_{22} \in A/P_i$ such that the matrix $M_i = \begin{pmatrix} \overline{b}_{11} & \overline{b}_{12} \\ \overline{b}_{21} & \overline{b}_{22} \end{pmatrix}$ is right and left invertible in the ring of all 2×2 matrices over A/P_i and the product $(h_i(a_1), h_i(a_2))M_i$ is a row of length 2 with zero second element $h_i(0)$). By **b**, there exist elements $b_{11}, b_{12}, b_{21}, b_{22} \in A$ such that the matrix $M = \begin{pmatrix} b_{11} & b_{12} \\ b_{21} & b_{22} \end{pmatrix}$ is right invertible in the ring of all 2×2 matrices over A and the product $(a_1, a_2)M$ is a row of length 2 with zero second element 0. Since all the matrices M_i are left invertible in the ring of all 2×2 matrices over A/P_i, it follows from 1) that the right invertible matrix M is also left invertible. Therefore, A is a right Hermite ring.

d. The proof of **d** is similar to the proof of **c**.

e. By **c**, A is a Hermite ring. By **d**, every square matrix over A is diagonalizable. By 4.1.6(b), A is a diagonalizable ring.

f. Every semilocal right Bezout ring is a right Hermite [193]. Therefore, every Pierce stalk of the ring A is a right Hermite ring. By **c**, A is a right Hermite ring.

g. Since every semilocal Bezout ring is a Hermite ring [193] and every square matrix over serial ring is diagonalizable [120], it follows from **e** that A is a diagonalizable ring.

h. The assertion follows from **g** and the property that every factor ring of a Bezout ring is a Bezout ring.

i. By assumption, every Pierce stalk A/P_i of the ring A is a right Bezout domain. By 4.1.5, every right Bezout domain is a right Hermite ring. Therefore, every Pierce stalk A/P_i of the ring A is a right Hermite ring. By **c**, A is a right Hermite ring.

j. Let $a \in A$. By assumption, aA is a flat right A-module. Since the module aA_A is isomorphic to the factor module $A_A/r(a)$ of the free module A_A and $r(A)$ is a finitely generated right A-module, aA is a finite-presented flat A-module. Since all finite-presented flat modules are projective, the module aA_A is projective and A is a right PP ring. By 3.2.11, A is a PP ring. By 4.2.2(b), every Pierce stalk of the ring A is a domain. □

4.2.7 Theorem. Let A be a reduced ring such that the right annihilator of each element of A is a finitely generated right ideal.
Then the ring A is a right Bezout ring if and only if A is a right Hermite ring.
Under these conditions, A is a right semihereditary, right stably finite ring which has the strongly regular right classical ring of fractions Q and contains all idempotents of the ring Q.
In addition, every finitely generated right ideal B of the ring A is a quasiprojective right A-module and a free cyclic right $A/r(B)$-module.

Proof. If A is a right Hermite ring, then A is a right Bezout ring by 2.2.3(b).

Let A be a right Bezout ring. By 3.2.6(d), A is a PF ring. By 4.2.6(j), A is a PP ring and every Pierce stalk of the ring A is a domain. Since the domain A/P is a factor ring of the right Bezout ring A, we have that A/P is a right Bezout domain. By 4.1.5, A/P is the right Hermite ring. By 4.2.6(i), A is a right Hermite ring.

Let 1) and 2) hold. By 4.2.6(j), A is a right PP ring and a right Bezout ring. Therefore, A is a right semihereditary ring. By 2.3.40(3),(4), A is a stably finite ring that has the strongly regular right classical ring of fractions. By 4.1.1(a), A contains all idempotents of the ring Q. By 3.2.6(e), every finitely generated right ideal B of the ring A is a quasiprojective right A-module and a free cyclic right $A/r(B)$-module. □

4.2.8 The completion of the proof of Theorems 4A and 4B. Theorem 4A follows from Theorem 4.2.7.
Theorem 4B follows from 4.2.6(h).

4.2.9 Open question. Is it true that every commutative Bezout domain is diagonalizable?

5 Bezout rings, Krull dimension

The main results of this section are Theorems 5A, 5B and 5C.

5A Theorem (Tuganbaev [183]). If A is a Bezout exchange ring without noncentral idempotents, then A is a diagonalizable ring.

5B Theorem (Tuganbaev [187]). If A is a right invariant, right Bezout exchange ring, then $B + r(X) = r(X/XB)$ for every finitely generated right A-module X and each ideal B of the ring A.

5C Theorem (Tuganbaev [188]). If A is a commutative arithmetical ring, then A has the Krull dimension if and only if every factor ring of the ring A is finite-dimensional and does not have idempotent proper essential ideals.

The comment to Theorem 5C are given below in 5.2.2.

Remark. The completion of the proof of Theorems 5A and 5B and 5C is given in 5.1.12, 5.1.13 and 5.2.12.

5.1 Bezout rings and modules

5.1.1 Bezout modules. Let A be a ring and M a Bezout right A-module.

a. If the ring A is right quasi-invariant, then the module M is distributive.

b. If the ring A is local, then M is a uniserial module.

Proof. **a.** We assume that there exists a nondistributive Bezout module M. By 1.1.5, there exists a 2-generated submodule X of the module M, which has a factor module $S \oplus T$ such that S and T are isomorphic simple modules. Since M is a Bezout module, $S \oplus T$ is a cyclic module. Therefore, there exist two distinct maximal right ideals B and C of the ring A such that $A/B \cong S \cong T \cong A/C$. By assumption, B and C are ideals. Therefore, $B = r(A/B)$ and $C = r(A/C)$. Since the annihilators of isomorphic modules coincide and $(A/B)_A \cong (A/C)_A$, we have $B = C$. This is a contradiction.

b. Since the local ring A is quasi-invariant, the module M is distributive by **a.** By 1.2.2(e), M is a uniserial module. □

5.1.2. If A is a left quasi-invariant ring, then each of its right or left invertible elements is invertible.

Proof. Let $a, b \in A$ and $ab = 1$. It is sufficient to prove that $Aa = A$.

We assume that $Aa \neq A$. Then a is contained in some maximal left ideal M. Since by assumption M is an ideal, $1 = ab \in M$. This is a contradiction. □

https://doi.org/10.1515/9783110659825-005

5.1.3. Let A be a right Bezout ring.

a. The ring A is right quasi-invariant if and only if A is right distributive.

b. If the ring A is left quasi-invariant, then A is a right distributive, right quasi-invariant ring.

Proof. **a.** If A is right quasi-invariant, then A is right distributive by 5.1.1(a). If A is right distributive, then A is right quasi-invariant by 1.2.3(d).

b. By **a**, it is sufficient to prove that any maximal right ideal M of the left quasi-invariant, right Bezout ring A is an ideal.

We assume the contrary. Then $A = AM$ and $1 = \sum_{i=1}^{n} a_i m_i$, where $a_i \in A$ and $m_i \in M$. Since A is a right Bezout ring, there exist elements $b_1, \dots, b_n, c_1, \dots, c_n \in A$ such that $\sum_{i=1}^{n} m_i c_i \equiv m \in M$ and $m_i = mb_i$ for all i. Since $m \in M$, we have $mA \neq A$. By 5.1.2, $Am \neq A$. Therefore, m is contained in some maximal left ideal B. By assumption, B is an ideal. Therefore, $1 = \sum_{i=1}^{n} a_i m b_i \in B$; this is a contradiction. □

5.1.4. If A is a Bezout ring, then the following conditions are equivalent.

1) A is right distributive.

2) A is left distributive.

3) A is right quasi-invariant.

4) A is left quasi-invariant.

5.1.4 follows from 5.1.3.

5.1.5. A ring A is a right Bezout ring if and only if all Pierce stalks of the ring A are right Bezout rings.

Proof. Since all factor rings of a right Bezout ring are right Bezout rings, every Pierce stalk of the ring A is a right Bezout ring.

We assume that every Pierce stalk of the ring A is a right Bezout ring. Let $m, n \in A$ and let $X_1 - (X_1Y_1 + X_2Y_2)Y_3$, $X_2 - (X_1Y_1 + X_2Y_2)Y_4$ be polynomials in noncommuting variables $X_1, X_2, Y_1, Y_2, Y_3, Y_4$. In addition, let A/P be a Pierce stalk of the ring A. Let $h\colon A \to A/P$ be the natural epimorphism. Since $h(A)$ is a right Bezout ring, there exist elements $\overline{a}, \overline{b}, \overline{c}, \overline{d} \in h(A)$ such that

$$0 = h(m) - (h(m)\overline{a} + h(n)\overline{b})\overline{c}\,, \qquad 0 = h(n) - (h(m)\overline{a} + h(n)\overline{b})\overline{d}\,.$$

By 4.2.5, there exist elements $a, b, c, d \in A$ such that $0 = m - (ma + nb)c$ and $0 = n - (ma + nb)d$. Therefore, A is a right Bezout ring. □

5.1.6. A ring A is an exchange ring if and only if all Pierce stalks of the ring A are exchange rings;

Proof. If A is an exchange ring, then each of its factor rings is an exchange ring.

We assume that all Pierce stalks of the ring A are exchange rings.

Let X_1, X_2, Y_1, Y_2 be noncommuting variables. We consider polynomials $f_1(X_1, Y_1, Y_2) = Y_1 - X_1Y_2$, $f_2(X_1, X_2, Y_1, Y_2) = X_2 - Y_1 - (X_2 - X_1)Y_2$ and $f_3(Y_1) = Y_1 - Y_1{}^2$. In addition, let A/P be a Pierce stalk of the ring A, $h\colon A \to A/P$ be the natural epimorphism and let $a \in A$. Since $h(A)$ is an exchange ring, there exist elements $\overline{e}, \overline{b}, \overline{c} \in h(A)$ such that $0 = \overline{e} - h(a)\overline{b} = f_1(h(a), \overline{e}, \overline{b})$, $0 = h(1) - \overline{e} - (h(1) - h(a))\overline{c} = f_2(h(a), h(1), \overline{e}, \overline{c})$, $0 = \overline{(e)} - \overline{e}^2 = f_3(\overline{e})$. By 4.2.5, there exist elements $e, b, c \in A$ such that $0 = e - ab = f_1(a, e, b)$, $0 = 1 - e - (1 - a)c = f_2(a, 1, e, c)$, $0 = e - e^2 = f_3(e)$. Then e is an idempotent, $e \in aA$ and $1 - e \in (1 - a)A$. Therefore, A is an exchange ring. □

5.1.7. For a ring A, the following conditions are equivalent.

1) All idempotents of the ring A are central.

2) For every proper ideal P of the ring A generated by central idempotents, all idempotents of the ring A/P are central.

3) Each Pierce stalk of the ring A does not have nontrivial idempotents.

Proof. 1) ⇒ 2). Let $h\colon A \to A/P$ be the natural epimorphism and $\overline{e} = \overline{e}^2 \in h(A)$. By 4.2.1(c), $\overline{e} = h(e)$ for some idempotent $e \in A$. By assumption, e is a central idempotent of the ring A. Then $h(e)$ is a central idempotent of the ring $h(A)$.

2) ⇒ 3). Let A/P be a Pierce stalk, $h\colon A \to A/P$ be the natural epimorphism, and let $\overline{e}$ be a nonzero idempotent of the ring $h(A)$. By 4.2.1(c), there exists an idempotent $e \in A$ such that $h(e) = \overline{e} \neq 0$. Since all idempotents of the ring A are central and $h(e) \neq 0$, the idempotent e is central in A and $e \notin P$. By 4.2.1(d), $1 - e \in P$. Then $h(e) = h(1)$.

3) ⇒ 1). Let $e = e^2 \in A$, $a \in A$, A/P be a Pierce stalk, and let $h\colon A \to A/P$ be the natural epimorphism. By assumption, $h(e)$ is a central idempotent of the ring $h(A)$. Therefore, $h(ea - ae) = 0$ and $ea - ae \in P$. By 4.2.2(b), A is the subdirect product of their Pierce stalks. Therefore, $ea - ae = 0$ and e is a central idempotent. □

5.1.8. For a ring A, the following conditions are equivalent.

1) A is an exchange ring without noncentral idempotents.

2) Every element of the ring A is a sum of an invertible element and central idempotent.

3) All Pierce stalks of the ring A are local rings.

Proof. 1) ⇒ 2). Let $x \in A$. Since A is an exchange ring without noncentral idempotents, there exist a central idempotent $e \in xA$ and elements $a, b \in A$ such that $e = xa$ and $1 - e = (1 - x)b$. Without loss of generality, we can assume that $a = ea = ae$ and $b = (1 - e)b = (1 - e)b$. Then ax and $b(1 - x)$ are central idempotents and

$$ax = a(xa)x = xa(ax) = xaxa = xa = e\,.$$

It can be similarly proved that $(1-x)b = 1-e$. In addition,

$$\begin{aligned}(a-b) &= xa + (1-x)b - b - a + ea + (1-e)b \\ &= e + (1-e) - b + b = 1\,.\end{aligned}$$

It can be similarly proved that $(a-b)[x-(1-e)] = 1$. Therefore, $x-(1-e)$ is an invertible element and $x = x-(1-e)+(1-e)$.

2) ⇒ 3). Let A/P be a Pierce stalk, $h\colon A \to A/P$ be the natural epimorphism, and let $h(a)$ be any noninvertible element in $h(A)$, where $a \in A$. It is sufficient to prove that $h(1)-h(a)$ is an invertible element in $h(A)$. By assumption, there exists a central idempotent $e \in A$ such that $e-a \in U(A)$. Then $h(e)-h(a) \in U(h(A))$. Since $h(a) \notin U(h(A))$, we have $h(e) \neq 0$. Then $e \in A\setminus P$ and the ideal $P+eA$ is generated by a central idempotent and properly contains ideal P, where A/P is a Pierce stalk. Therefore, $P+eA = A$. Then $h(e) = h(1)$. Therefore, $h(1)-h(a) \in U(h(A))$.

3) ⇒ 1). Since any local ring does not have noncentral idempotents, it follows from 5.1.7 that all idempotents of the ring A are central. In addition, every local ring is an exchange ring. By 5.1.6, A is an exchange ring. □

5.1.9. A ring A is right distributive if and only if all Pierce stalks of the ring A are right distributive.

Proof. If the ring A is right distributive, then all its factor rings are right distributive.

Let all Pierce stalks of the ring A be right distributive and let Y_1+Y_2-1, $X_1Y_1-X_2Y_3$ and $X_2Y_2-X_1Y_4$ be polynomials in noncommuting variables $X_1, X_2, Y_1, Y_2, Y_3, Y_4$. In addition, let A/P be a Pierce stalk, $h\colon A \to A/P$ be the natural epimorphism, and let $m, n \in A$. Since the ring $h(A)$ is right distributive, it follows from 1.1.2 that there exist elements $\overline{a}, \overline{b}, \overline{c}, \overline{d} \in h(A)$ that

$$\overline{a}+\overline{b}-h(1) = 0\,, \quad h(m)\overline{a}-h(n)\overline{c} = 0\,, \quad h(n)\overline{b}-h(n)\overline{d} = 0\,.$$

By 4.2.5, there exist elements $a, b, c, d \in A$ such that $0 = a+b-1$, $0 = ma-nc$ and $0 = nb-md$. By 1.1.2, A is right distributive. □

5.1.10 Theorem. For a ring A, the following conditions are equivalent.

1) A is an exchange right Bezout ring without noncentral idempotents.

2) A is a right quasi-invariant exchange ring that is a right Bezout ring.

3) A is a right distributive exchange ring.

4) Every Pierce stalk A/P of the ring A is a right uniserial ring.

Proof. 1) ⇒ 4). By 5.1.8, every Pierce stalk A/P is a local ring. Since all factor rings of a right Bezout ring are right Bezout rings, every Pierce stalk A/P is a local right Bezout ring. By 5.1.1(b), A/P is a right uniserial ring.

4) ⇒ 1), 4) ⇒ 2) and 4) ⇒ 3). By 5.1.9, the ring A is right distributive. Since every right uniserial ring is a right Bezout exchange ring, every Pierce stalk of the ring A is an exchange right Bezout ring. By 5.1.5 and 5.1.8, A is an exchange right Bezout ring without noncentral idempotents. By 1.2.3(d), the ring A is right quasi-invariant.

2) ⇒ 3). The assertion follows from 5.1.1(a).

3) ⇒ 4). By 1.1.7(a), all idempotents of the ring A are central. Since all factor rings of right distributive rings are right distributive, every Pierce stalk A/P is a right distributive local ring by 4.2.3(c). By 1.2.2(d), A/P is a right uniserial ring. □

5.1.11 Theorem. For a ring A, the following conditions are equivalent.

1) A is a diagonalizable exchange ring without noncentral idempotents.

2) A is a Bezout exchange ring without noncentral idempotents.

3) A is a quasi-invariant exchange Bezout ring.

4) A is a distributive exchange ring.

5) All Pierce stalks of the ring A are uniserial rings.

Proof. The equivalence of conditions 2), 3), 4) and 5) follows from 5.1.10.

1) ⇒ 2). Since every diagonalizable ring is a Hermite ring, 2) follows from 2.2.3(b).

2) ⇒ 1). By 2), A is an exchange ring without noncentral idempotents. Since 2) and 5) are equivalent, every Pierce stalk of the ring A is a uniserial ring. In particular, every Pierce stalk of the ring A is a serial Bezout ring. By 4.2.6(h), A is a diagonalizable ring. □

5.1.12 The completion of the proof of Theorems 5A. Theorem 5A follows from Theorem 5.1.11.

5.1.13 The completion of the proof of Theorem 5B. Since A is a right invariant a Bezout exchange ring, it follows from Theorem 5.1.11 that A is a diagonalizable ring. By Theorem 2B from the beginning of Section 2, $B + r(X) = r(X/XB)$ for every finitely generated right A-module X and each ideal B of the ring A.

5.2 Rings with Krull dimension

5.2.1 Krull dimension of modules and rings. We recall the transfinite definition of the Krull dimension Kdim M of the module M, see[82]. (We remark that there are modules without Krull dimension.)
By definition, it is assumed that the zero module has the Krull dimension −1 and the Krull dimension of each nonzero Artinian module is zero.
We assume that α is an ordinal > 0, modules with Krull dimension β defined for all ordinals $\beta < \alpha$, and M is a module such that Kdim $M \neq \beta$.

One says that the *Krull dimension* Kdim M of the module M is equal to α if for any infinite properly descending chain of $M_1 > M_2 > \dots$ of submodules in M, there exists a positive integer n such that $\mathrm{Kdim}(M_n/M_{n+1}) < \alpha$.

If A is a ring and the module A_A has Krull dimension, then the dimension is called the *right Krull dimension* Kdim A_A of the ring A.
For example, the Krull dimension of the residue ring $\mathbb{Z}/n\mathbb{Z}$ is equal to 0, and the Krull dimension of the polynomial ring $k[x_1, \dots, x_n]$ over any field k is equal to n.

a. All factor modules and submodules of the module with Krull dimension have Krull dimension, every extension of the module with Krull dimension with the use of the module with Krull dimension has Krull dimension and every Noetherian module has Krull dimension; see [82].

b. The class of all rings with right Krull dimension is larger than the class of all right Noetherian rings and the rings with right Krull dimension have many useful properties of right Noetherian rings.
For example, if A is a ring with right Krull dimension, then its prime radical P is nilpotent and the factor ring A/P has the semisimple Artinian right classical ring of fractions; see [82].

c. Any module with Krull dimension is finite-dimensional; see [82, Proposition 1.4].

5.2.2 Arithmetical rings and Krull dimension. An ideal B is called an *idempotent* ideal if $B = B^2$.
In [118], it is proved that commutative uniserial ring A has Krull dimension if and only if A does not have idempotent proper nonzero ideals.
In this case, it is clear that every factor ring of the ring A does not have idempotent proper nonzero ideals.
In addition, every factor ring of any commutative uniserial ring is a finite-dimensional uniserial ring such that each of its nonzero ideals is an essential.

a. We remark that $\mathbb{Z}$ is a commutative arithmetical ring with Krull dimension 1. In addition, $\mathbb{Z}$ does not have idempotent proper nonzero ideals and its factor ring $\mathbb{Z}/6\mathbb{Z}$ has idempotent proper nonzero ideals.

b. The direct product of $A = \prod_{i\in I} A_i$ of any set of fields A_i is a commutative arithmetical ring such that all its ideals are idempotent ideals.
In this case, the ring A has Krull dimension if and only if the set I is finite.

c. Let A be a commutative ring with Krull dimension. In [117, Théorème 12], it is proven that $A = A_1 \times \dots \times A_n$, where each of the rings A_i have Krull dimension and do not have idempotent proper nonzero ideals. Since every factor ring of the ring A has Krull dimension, every factor ring $\overline{A}$ of the ring A does not have idempotent proper essential ideals. In addition, $\overline{A}$ is a finite-dimensional ring.

5.2.3. Let A be a right distributive ring.

a. If $P_1 \subset P_2 \subset \dots$ is an infinite properly ascending chain of completely prime right ideals of the ring A, then $\cup_{i=1}^{\infty} P_i$ is an idempotent proper completely prime right ideal.

b. If the ring A does not have idempotent proper completely prime ideals, then A is a ring with the maximum condition on completely prime ideals.

c. If the ring A does not have idempotent proper completely prime right ideals, then A is a ring with the maximum condition on completely prime right ideals.

d. If A is a right finite-dimensional reduced ring, then A is a finite direct product of right distributive right uniform domains.

Proof. **a.** We denote by X the right ideal $\cup_{i=1}^{\infty} P_i$. Since all P_i are proper right ideals, X is a proper right ideal. Since all P_i are completely prime right ideals, X is a completely prime right ideal.

We assume that $X \neq X^2$. Let $x \in X \setminus X^2$. Then $x \in P_i$ for some i. Since P_i is a completely prime right ideal and X properly contains P_i, we have that $X^2 \not\subseteq P_i$ and there exists an element $y \in X^2 \setminus P_i$. By 1.1.2, there exist two elements $a, b \in A$ such that $a + b = 1$, $xa \in yA \subseteq X^2$ and $yb \in xA \subseteq P_i$. Since $x \notin X^2$ and $xa \in X^2$, we have $xb = x - xa \notin X^2$. Therefore, $b \notin X$. Since $y \notin P_i$ and the element yb is contained in the completely prime right ideal P_i, we have $b \in P_i \subseteq X$. This is a contradiction.

b, c. The assertions follow from **a**.

d. The right finite-dimensional ring A contains an essential right ideal $x_1A \oplus \cdots \oplus x_nA$, where each of the right ideals x_iA are uniform. Since the ring A is right distributive, it follows from 1.2.2(d) that $x_1A \oplus \cdots \oplus x_nA = xA$, where $x = x_1 + \cdots + x_n$. Since A is a reduced ring and xA is an essential right ideal, it follows from 1.3.1(b) that $r(x) = 0$. Therefore, xA_A is a free cyclic module. In addition, $xA = (xA \cap x_1A) \oplus \cdots \oplus (xA \cap x_nA)$. Therefore, A is a finite direct product of right distributive right uniform domains. □

5.2.4. Let A be a right invariant arithmetical ring and N be the set of all right or left zero divisors of the ring A.

a. If A does not have idempotent proper prime ideals, then A is a ring with the maximum condition on prime ideals.

b. If the set N is a prime ideal of the ring A, then A has the right uniserial right classical ring of fractions Q, $NQ = J(Q)$ and the ring A is right uniform.

Proof. Since A is a right invariant arithmetical ring, A is a right distributive ring that has the right classical ring of fractions Q.

a. Since every prime ideal of the right invariant ring A is completely prime, the assertion follows from 5.2.3(b).

b. Since the ring A is right invariant, the prime ideal N is completely prime and the set S of all nonzero divisors of the ring A has the form $A \setminus N$. By 1.3.2(k), Q is a right uniserial ring, $J(Q) = NQ$ and the ring A is right uniform. □

5.2.5. Let A be a right invariant arithmetical ring such that all its right zero divisors are left zero divisors and the intersection of any two nonzero ideals is equal to zero. Let N be the set of all right or left zero divisors of the ring A.

a. N is a completely prime ideal of the ring A, the ring A has the right uniserial right classical ring of fractions Q, and $NQ = J(Q)$.

b. If A has a completely prime nil ideal P, then $P = xP$ for any nonzero divisor x of the ring A, $P = QP$ is a left ideal of the ring Q, $PQ = QPQ$ is an ideal of the ring Q and $(PQ)^n = P^n Q$ for every positive integer n.

Proof. Since A is a right invariant arithmetical ring such that the intersection of any two its nonzero ideals is equal to zero, A is a right distributive right uniform ring.

a. Since the ring A is right uniform, the set N_1 of all its left zero divisors coincides with the set of all elements of the ring A whose right annihilators are essential right ideals. In this case, it is well known that N_1 is a completely prime ideal of the ring A; e.g., see [166, 5.30(2)] or [169, Lemma 1.4]. Since all right zero divisors of the ring A are left zero divisors, $N = N_1$. Therefore, N is a completely prime ideal. By 5.2.4(b), A has the right uniserial right classical ring of fractions Q and $NQ = J(Q)$.

b. Let $y \in P$. By 1.1.2, there exist elements $a, b \in A$ such that $a + b = 1$ and $xa, yb \in xA \cap yA$. Since $x \notin P$ and xa is an element of the completely prime ideal P, we have that a is an element of the nil ideal P. Then $b = 1 - a$ is an invertible element and $yb \in xA$. Therefore, $y = ybb^{-1} \in xA$ and $P = xP$. Since x is an invertible element of the ring Q and $P = xP$, we have $P = x^{-1}P$. Therefore, $P = QP$. Then $PQ = QPQ$ is an ideal of the ring A and $(PQ)^n = P^n Q$ for every positive integer n. □

5.2.6. Let A be a commutative arithmetical uniform ring, P be the prime radical of the ring A, and let A/P be a finite-dimensional ring.

a. P is a completely prime nil ideal and either $P = P^2$ or P is a nilpotent ideal.

b. If the ring A does not have idempotent proper essential ideals, then P is a nilpotent ideal.

Proof. **a.** Since A/P is a commutative finite-dimensional distributive reduced ring, A/P is a finite direct product of domains by 5.2.3(d). Since P is a nil ideal, all idempotents of the ring A/P can be lifted to idempotents of the commutative indecomposable ring A. Therefore, the ring A/P does not have nontrivial idempotents. Then A/P is a domain and P is a completely prime nil ideal. By 5.2.5, the ring A has the uniserial classical ring of fractions Q, $P = xP$ for any nonzero divisor x of the ring A, $P = QP$ is an ideal of the ring Q and $P^n = P^n Q$ for every positive integer n.

We assume that $P \neq P^2$. Let $p \in P \setminus P^2$. Then pQ is a nilpotent ideal of the commutative uniserial ring Q that is not contained in the ideal $P^2 = P^2Q$. Then the ideal P^2 is contained in the nilpotent ideal pQ. Therefore, the ideal P^2 is nilpotent. Then P is a nilpotent ideal.

2) Without loss of generality, we can assume that P is a nonzero proper ideal of the ring A. Since A is a uniform ring, P is an essential ideal. By assumption, the ring A does not have proper essential idempotent ideals. Therefore, $P \neq P^2$. By **a**, P is a nilpotent ideal. □

5.2.7 ([118, Proposition 2]). Let M be a module such that all factor modules of the module M are finite-dimensional and any nonzero factor module Q of the module M contains a nonzero submodule with Krull dimension. Then M has Krull dimension.

5.2.8 ([82, 1.4, 7.1, 7.3, 7.4]). Let A be a ring that has right Krull dimension. Then every factor ring of the ring A is right finite-dimensional, A is a ring with the maximum condition on prime ideals, and for every proper ideal B of the ring A, there exist prime ideals $P_1, \dots, P_n$ of the ring A such that $P_1 \cdot \dots \cdot P_n \subseteq B$ and each of the ideals P_i contains the ideal B.

5.2.9. Let A be a ring such that all cyclic right A-modules are finite-dimensional.

a. If there exists a finite set $\{P_1, \dots, P_n\}$ ideals of the ring A such that each of the cyclic right A-modules A/P_i has Krull dimension, then the ring $A/(P_1 \cdot \dots \cdot P_n)$ has right Krull dimension.

b. We assume that there exists a set $\mathfrak{Q}$ of proper ideals of the ring A such that ascending chains of ideals from $\mathfrak{Q}$ are stabilized at a finite step and for any nonzero right ideal B of the ring A, there exist ideals $P_1, \dots, P_n \in \mathfrak{Q}$ such that

$$P_1 \cdot \dots \cdot P_n \subseteq B \subseteq P_1 \cap \dots \cap P_n .$$

Then A has right Krull dimension.

Proof. **a.** V. T.Markov informed the author of the proof of this assertion. We use the induction by n. For $n = 1$, there is nothing to prove. Let $n > 1$ and $Q = P_1 \dots P_{n-1}$. By the induction hypothesis, the factor ring A/Q has right Krull dimension. Let X be an arbitrary nonzero factor module of the right A-module Q/QP_n. Then X is a submodule of the right A/P_n-module Q/QP_n. Since X is a right module over the ring A/P_n with right Krull dimension, every nonzero submodule of the A/P_n-module X contains a nonzero submodule with Krull dimension. Then every nonzero submodule of the A-module X contains a nonzero submodule with Krull dimension. In addition, X_A is a submodule of the cyclic A-module A/QP_n and all factor modules of the cyclic A-module A/QP_n are finite-dimensional by assumption. Therefore, all factor modules of the module X_A are finite-dimensional. By 5.2.7, the module X_A has Krull dimension. Since X is an arbitrary nonzero factor module $(Q/QP_n)_A$, the module $(Q/QP_n)_A$ has Krull dimension. In addition, $(A/Q)_A$ has Krull dimension by the induction hypothesis. Since $(A/QP_n)_A$ is an extension of the module $(Q/QP_n)_A$ with Krull dimension with the use of the module $(A/Q)_n$, we have that $(A/QP_n)_A$ has Krull dimension. Therefore, the ring A/QP_n has right Krull dimension.

b. We assume that A does not have right Krull dimension. We construct a properly ascending sequence of ideals $Q_0 \subset Q_1 \subset Q_2 \dots$ in $\mathfrak{Q} \cup \{0\}$ such that the ring A/Q_i does not have Krull dimension for all $i \geq 0$. We set $Q_0 = 0$. We assume that the ideals $Q_0 \subset \cdots \subset Q_k$ have been constructed, where $k \geq 0$. Since the right A-module A/Q_k does not have Krull dimension, there exists a right ideal B of the ring A such that $B \supseteq Q_k$, $B/Q_k \neq 0$, and $(A/B)_A$ does not have Krull dimension. By assumption, there exist ideals $P_1, \dots, P_n \in \mathfrak{Q}$ such that

$$P_1 \cdot \cdots \cdot P_n \subseteq B \subseteq P_1 \cap \cdots \cap P_n \,.$$

By 5.2.7, there exists an integer $i \in \{1, \dots, n\}$ such that the cyclic right A-module A/P_i does not have Krull dimension. We set $Q_{k+1} = P_i$. We have an infinite properly ascending chain $Q_1 \subset Q_2 \subset \dots$ of ideals from $\mathfrak{Q}$; this contradicts the assumption. □

5.2.10. For a right invariant ring A, the following conditions are equivalent.

1) The ring A has right Krull dimension.

2) A is a ring with the maximum condition on prime ideals, all factor rings of the ring A are right finite-dimensional and for any nonzero ideal B of the ring A, there exist ideals $P_1, \dots, P_n \in \mathfrak{Q}$ such that

$$P_1 \cdot \cdots \cdot P_n \subseteq B \subseteq P_1 \cap \cdots \cap P_n \,.$$

Proof. The implication 1) ⇒ 2) follows from 5.2.8.

2) ⇒ 1). The ring A is right invariant and all factor rings of the ring A are right finite-dimensional. Therefore, all cyclic right A-modules are finite-dimensional. We denote by $\mathfrak{Q}$ the set of all prime ideals of the ring A. Now the assertion follows from 5.2.9(2). □

5.2.11. Let A be a commutative arithmetical ring and every factor ring of the ring A is finite-dimensional and does not have idempotent proper essential ideals. Then A has Krull dimension.

Proof. By assumption, every factor ring of the ring A is finite-dimensional. By 5.2.4, A is a ring with the maximum condition on prime ideals. Let B be an arbitrary proper ideal of the ring A. By 5.2.10, it is sufficient to prove the existence of prime ideals $Q_1, \dots, Q_n$ of the ring A such that $Q_1 \cdot \cdots \cdot Q_n \subseteq B$ and each of the ideals Q_i contains ideal B.

By assumption, A/B is a finite-dimensional ring. Therefore, there exist ideals $B_1, \dots, B_k$ of the ring A such that $B_1 \cap \cdots \cap B_k = B$ and every factor ring A/B_i is a uniform ring. Let $P_1, \dots, P_k$ be ideals of the ring A such that $B_i \subseteq P_i$ and let P_i/B_i be the prime radical of the ring A/B_i, $i = 1, \dots, k$. By assumption, every factor ring A/P_i is a uniform ring. By assumption, every factor ring A/B_i is a commutative arithmetical ring without idempotent proper essential ideals. By 5.2.6, P_i/B_i is a prime nilpotent ideal of the ring A/B_i, $i = 1, \dots, k$. Therefore, there exist positive integers $n_1, \dots, n_k$ such that $P_i^{n_i} \subseteq B_i$, $i = 1, \dots, k$. Therefore, the ideal $X = P_1^{n_1} \cdot \dots \cdot P_k^{n_k}$ is contained in the ideal $B_1 \cap \dots \cap B_k = B$ and every prime ideal P_i contains ideal B. This implies the required assertion. □

Proof.

5.2.12 The completion of the proof of Theorem 5C. 1) ⇒ 2). The assertion follows from 5.2.2(c).

2) ⇒ 1). The assertion has been proven in 5.2.11. □

5.2.13 Open question. Let A be an invariant arithmetical ring and every factor ring of A is finite-dimensional and does not have idempotent proper essential ideals. Is it true that A has right Krull dimension?

Part II: EXTENSION OF AUTOMORPHISMS AND ENDOMORPHISMS

6 Semi-Artinian and nonsingular modules

The main results of this section are Theorems 6A, 6B and 6C.

6A Theorem (Tuganbaev [184]). If M is a semi-Artinian[1] module, then M is an automorphism-extendable module if and only if M is an automorphism-invariant module.

6B Theorem (Tuganbaev [174]). If M is a module over an Artinian serial ring, then M is an automorphism-extendable module if and only if M is a quasi-injective module.

6C Theorem (Tuganbaev [176]). Let $M = T \oplus U$, where T is an injective module, U is a nonsingular module, and $\mathrm{Hom}(T', U) = 0$ for any submodule T' of the module T. The module M is automorphism-extendable if and only if U is an automorphism-extendable module.

Remark. The completion of the proof of Theorems 6A, 6B and 6C is given in 6.2.10.

6.1 Semi-Artinian Modules

6.1.1 Free, injective and projective modules. Relative injectivity and projectivity. We recall that the module M is said to be *injective with respect to the module* X or X*-injective* if for any submodule X_1 in X, every homomorphism $X_1 \to M$ can be extended to a homomorphism $X \to M$.
A module is said to be *injective* if it is injective with respect to any of the module.
A module is said to be *quasi-injective* if it is injective with respect to itself.

Let A be a ring and let X, M be two right A-modules.

a. The module X is a free module of rank $\aleph$ if and only if X is isomorphic to the direct sum of some set I of cardinality $\aleph$ of isomorphic copies of the free cyclic module A_A. In particular, there exist free modules of any rank $\aleph$.

If X is a free module, then every mapping f from the basis $\{x_i\}_{i\in I}$ of the module X into the module M by the rule $g(\sum x_i a_i) = \sum f(x_i)a_i$ can be correctly extended to a homomorphism $g\colon X \to M$. In addition, if $\{f(x_i)\}_{i\in I}$ is a system of generators of the module M, then $g\colon X \to M$ is an epimorphism.

b. Every module with generator system of cardinality $\aleph$ is a homomorphic image of the free module of rank $\aleph$.

c. Every module is projective and injective with respect to any semisimple module.

---

1 A module M is said to be *semi-Artinian* if each of its nonzero factor modules has a simple submodule.

https://doi.org/10.1515/9783110659825-006

d. The class $\mathcal{X}$ of all modules X, such that the module M is X-injective, contains all submodules, homomorphic images and direct sums of modules from $\mathcal{X}$.

The class $\mathcal{Y}$ of all modules Y such that the module M is Y-projective contains all homomorphic images, submodules and finite direct sums of modules from $\mathcal{Y}$.

e. If the module X contains an isomorphic copy of the module A_A and the module M is injective with respect to the module X, then M is an injective module. Consequently, the ring A is right injective if and only if the ring A is right quasi-injective.

f. All direct summands and direct products of modules, which are injective with respect to the module M, are M-injective. In particular, all direct summands and direct products of injective modules are injective.

All direct summands and direct sums of modules, which are projective with respect to the module M, are M-projective. In particular, all direct summands and direct sums of projective modules are projective.

g. If the module M is injective with respect to the module X and there exists a monomorphism $f\colon M \to X$, then $f(M)$ is a direct summand of X, M is quasi-injective, and M is isomorphic to a direct summand of the module X. In particular, if either the module X is indecomposable or $f(M)$ is an essential submodule in X, then $f\colon M \to X$ is an isomorphism.

h. If the module M is projective with respect to the module X and there exists an epimorphism $h\colon X \to M$, then $\operatorname{Ker} h$ is a direct summand of X and M is a quasiprojective module which is isomorphic to the to a direct summand of the module X. In particular, if the module X is indecomposable, then $h\colon X \to M$ is an isomorphism.

i. If Y is a submodule of the module X and the module X/Y is projective with respect to X, then Y is a direct summand of X. In addition, the projectivity of the cyclic module xA is equivalent to each of the following conditions: 1) $r(x)$ is a direct summand of A_A; 2) $r(x) = eA$ for some idempotent $e \in A$.

j. Every free module is projective.
The module M is projective if and only if M is isomorphic to a direct summand of a free module.
A module M is a projective if and only if for every module epimorphism $h\colon X \to M$, the module $\operatorname{Ker} h$ is a direct summand of X.

k. If $M = \oplus_{i\in I} M_i$ and for any $i \in I$, the modules M_i and $\oplus_{j\neq i} M_j$ are M_i-injective, then M is a quasi-injective module.

Proof. **a.** The assertion is directly verified.

b. Let $\{m_i\}_{i\in I}$ be a generator system of cardinality $\aleph$ of the module M_A. We take a free module X_A with basis $\{x_i\}_{i\in I}$ of cardinality $\aleph$. We define a mapping $f\colon \{x_i\}_{i\in I} \to \{m_i\}_{i\in I}$ such that $f(x_i) = m_i$ for all $i \in I$. By **a**, f can be extended to a module epimorphism $X \to M$.

c. The assertion follows from the property that any submodule of an arbitrary semisimple module X is a direct summand in X.

d. We prove only the first assertion, since the second assertion is similarly proven. If M is X-injective, then it is directly verified that M is injective with respect to any submodule of the module X. We assume that $h\colon X \to \overline{X}$ is an epimorphism, $\overline{Y}$ is a submodule in $\overline{X}$, $\overline{g} \in \operatorname{Hom}(\overline{Y}, M)$. We denote by Y the complete pre-image of $\overline{Y}$ in X under the action of h. Let h_Y be the restriction of h to Y. By assumption, the homomorphism $\overline{g}h_Y\colon Y \to M$ can be extended to a homomorphism $f\colon X \to M$. Since $\operatorname{Ker} h \subseteq Y$, we have $\operatorname{Ker} h \subseteq \operatorname{Ker} f$. Therefore, f can be extended to some homomorphism $\overline{f}\colon \overline{X} \to M$ and M is $\overline{X}$-injective.

Let $\{X_i\}_{i\in I}$ be a set of modules such that M is X_i-injective for all $i \in I$. Let $Y = \oplus_{i\in I} X_i$, Y_1 be a submodule in Y, $f_1 \in \operatorname{Hom}(Y_1, M)$, $\mathcal{E}$ be the set of all pairs (L, f_L), where L is a submodule in Y which contains Y_1, and let f_L be a homomorphism from L into M extending f_1. We define a relation $\leq$ on $\mathcal{E}$ such that $(L, f_L) \leq (Q, f_Q)$ if and only if $L \subseteq Q$ and f_L can be extended to f_Q. We can verify that $\leq$ is a partial order on $\mathcal{E}$ and every nonempty chain in $\mathcal{E}$ has the upper bound. By the Zorn lemma, $\mathcal{E}$ contains a maximal element $(\overline{Y}, \overline{f})$. It is sufficient to prove that $\overline{Y} = Y$, which is equivalent to the inclusions $X_i \subseteq \overline{Y}$ for all $i \in I$. Since M is X_i-injective, the restriction of the homomorphism $\overline{f}$ to $X_i \cap \overline{Y}$ can be extended to some homomorphism $f_i\colon X_i \to M$. Let $u\colon (X_i + \overline{Y}) \to M$ be a homomorphism such that $u(x + y) = f_i(x) + \overline{f}(y)$ for all $x \in X_i$ and $y \in \overline{Y}$. This homomorphism is well-defined (if $x + y = 0$, then $x = -y \in X_i \cap \overline{Y}$ and $u(x + y) = \overline{f}(-y) + \overline{f}(y) = 0$). By construction, $X_i + \overline{Y} = \overline{Y}$. Therefore, $X_i \subseteq \overline{Y}$, which is required.

e. By **d**, M is injective with respect to the module A_A. In addition, any right A-module is a homomorphic image of the direct sum of some set of copies of the module A_A. Now the assertion follows from **d**.

f. We prove only the first assertion, since the second assertion is similarly proven. Let N be a right A-module and $N = \prod_{i\in I} N_i$. It is clear that the M-injectivity of the module N implies the M-injectivity of all modules N_i. Let's assume now that all modules N_i are M-injective. Let X be a submodule in M, $f \in \operatorname{Hom}(X, N)$, and let $\pi_i\colon N \to N_i$ be natural projections. All homomorphisms $\pi_i f\colon X \to N_i$ can be extended to homomorphisms $g_i\colon M \to N_i$ which define a natural extension of the homomorphism $g\colon M \to N$.

g. By **d**, M is an $f(M)$-injective module. Since $f(M) \cong M$, we have that M is quasi-injective. Since $f(M)$ is X-injective, a natural embedding $f(M) \to X$ can be extended to a homomorphism $g\colon X \to f(M)$. Then g is the projection X onto $f(M)$. Therefore, $f(M)$ is a direct summand of X.

h. By **d**, M is a quasiprojective module. Since M is an X-projective module, there exists a homomorphism $g\colon M \to X$ with $1_M = hg$. We set $\pi \equiv 1 - gh \in \operatorname{End}(X)$. Since

$\pi^2 = 1 - gh - gh + g(hg)h = 1 - gh = \pi$, we have $X = \pi(X) \oplus (1 - \pi)(X)$. Furthermore,

$$h\pi(X) = (h - hgh)(X) = (h - h)(X) = 0\,, \quad \pi(X) \subseteq \operatorname{Ker} h\,,$$
$$\operatorname{Ker} h = gh(\operatorname{Ker} h) + (1 - gh)(\operatorname{Ker} h) = \pi(\operatorname{Ker}(h)) \subseteq \pi(X)\,.$$

Therefore, $\operatorname{Ker} h = \pi(X)$ and $M \cong (1 - \pi)(X)$.

i. The first assertion follows from **h**. The second assertion follows from the first assertion and the property that for any cyclic module xA, there exists an isomorphism $f\colon xA \to A_A/r(x)$ such that $f(xa) = a + r(x)$ for all $a \in A$.

j. By **f**, it is sufficient to prove that A_A is a projective module. Let $h\colon X \to \bar{X}$ be an arbitrary epimorphism of right A-modules and let $\bar{f}\colon A_A \to \bar{X}$ be a homomorphism. There exists an element $x \in X$ such that $h(x) = \bar{f}(1)$. By **a**, the mapping $f(a) = xa$ is a correctly defined homomorphism from A_A into X. In addition, $hf = \bar{f}$.

k. The assertion is known; it is proven in [131, Proposition 1.18]. □

6.1.2 Injective hulls. Let M be a module. An *injective hull* of M is any injective module, which is an essential extension of the module M. The module M is often identified with its isomorphic copies; an injective hull of the module M is an injective module, which is an essential extension of an isomorphic copy of M.

a. Every module has an injective hull and all injective hulls of the module M are isomorphic to each other.

b. The module M is injective if and only if for any module monomorphism $f\colon M \to X$, the module $f(M)$ is a direct summand of X. In this case, f is an isomorphism if and only if $f(M)$ is an essential submodule in X.

c. If M is a quasi-injective (for example, injective) module, then the indecomposability of the module M is equivalent to the property that for any endomorphism f of the module M, at least one of the endomorphisms f, $1 - f$ is an automorphism, i.e., the endomorphism ring End M of the module M is a local ring. In this case, all submodules of the module M are uniform.

d. A module M is a quasi-injective if and only if M is an endomorphism-invariant module, i.e., M is a fully invariant submodule of its injective hull.

e. A module M_A is quasi-injective if and only if M is a quasi-injective $A/r(M)$-module, where $r(M)$ is an annihilator of the module M.

The assertions **a**–**e** from 6.1.2 are well known.

6.1.3. If A is a ring and M is a right A-module, then the following conditions are equivalent.

1) M is an automorphism-invariant module.

2) $M = \alpha(M) = \alpha^{-1}(M)$ for any an automorphism α of the injective hull E of M.

3) Every isomorphism between any two essential submodules of the module M can be extended to an endomorphism of the module M.

4) Every isomorphism between any essential submodules of the module M can be extended to an automorphism of the module M.

5) M is a characteristic submodule of some injective module Q.

Consequently, every pseudo-injective module is automorphism-invariant.

Proof. Let E be the injective hull of the module M.

The implications 2) $\Rightarrow$ 1), 5) $\Rightarrow$ 1) and 4) $\Rightarrow$ 3) are obvious.

3) $\Rightarrow$ 1). Let α be an automorphism of the module E, $X' = M \cap \alpha(M) \subseteq M$, $X = \alpha^{-1}(X') \subseteq M$. Since $\alpha(M)$ is an essential submodule of the module $E = \alpha(E)$, we have that X' is an essential submodule of the module M. In addition, $X = \alpha^{-1}(X')$ is an essential submodule of the module $E = \alpha^{-1}E$. Since α induces the isomorphism of the module X onto X', it follows from 3) that there exists an endomorphism β of the module M coinciding with α on X. We assume that $z \in M \cap (\alpha - \beta)(M)$. Then $z = (\alpha - \beta)(y)$, where $y \in M$. Then $\alpha(y) = \beta(y) - z \in M$. Therefore, $y \in X$. By construction, $(\alpha - \beta)(y) = z = 0$. Therefore, $M \cap (\alpha - \beta)(M) = 0$. Therefore, $(\alpha - \beta)(M) = 0$, since M is an essential submodule in E. Therefore, $\alpha(M) \subseteq M$.

1) $\Rightarrow$ 2). Since M is an automorphism-invariant module, $\alpha(M) \subseteq M$ and $\alpha^{-1}(M) \subseteq M$. Therefore, $M = \alpha(M) = \alpha^{-1}(M)$.

2) $\Rightarrow$ 4). Let X, X' be two essential submodules of the module M and let $\alpha' \colon X \to X'$ be an isomorphism. Since E is an injective module, there exists an endomorphism α of the module E, which coincides with α' on X. Since X is an essential submodule in E and $X \cap \operatorname{Ker} \alpha = 0$, we have that α is a monomorphism. Therefore, $\alpha(E) \cong E$ and $\alpha(E)$ is an injective module. Then $\alpha(E)$ is a direct summand of the module E. In addition, $\alpha(E)$ contains the submodule $\alpha(X)$ and $X' = \alpha X$. Therefore, $\alpha(E)$ is an essential direct summand of the module E. Then $\alpha(E) = E$ and α is an automorphism of the module E. By 2), the restriction α to M is an endomorphism of the module M which coincides with α' on X.

1) $\Rightarrow$ 5). Let $Q = E \oplus F$, where E is the injective hull of the module M. Every automorphism α of the module E can be extended to an automorphism β of the module Q with the use of the rule $\beta(x + y) = \alpha(x) + y$. By assumption, $\beta(M) \subseteq M$. Then $\alpha(M) = \beta(M) \subseteq M$.

1) $\Leftrightarrow$ 6). Since the equivalence of 1) $\Leftrightarrow$ 3) has been proven, the assertion is directly verified. □

6.1.4. Let Q be the injective hull of the module M, α be an endomorphism of the module Q, $X = \{m \in M \mid \alpha(m) \in M\}$, and let $f = \alpha|_X \colon X \to M$ be the restriction of the endomorphism α to the module X.

a. If the homomorphism $f =: X \to M$ can be extended to some endomorphism g of the module M, then $\alpha(M) \subseteq M$.

b. If $f(X) \subseteq X$ and the endomorphism f of the module X can be extended to some endomorphism g of the module M, then $\alpha(M) \subseteq M$.

c. If $\alpha = \alpha^2$ and every idempotent endomorphism of the module X can be extended to some endomorphism of the module M, then $\alpha(M) \subseteq M$.

Proof. **a.** Since Q is an injective module, the endomorphism g of the module M can be extended to some endomorphism β of the module Q.

We assume that $(\alpha - \beta)(M) = 0$. Then $\alpha(M) = \beta(M) \subseteq M$, which is required.

We assume that $(\alpha - \beta)(M) \neq 0$. Since Q is an essential extension of the module M and $X = \{m \in M \mid \alpha(m) \in M\}$, we have that X is an essential submodule in Q. Then $X \cap (\alpha - \beta)(M)$ is a nonzero submodule in M, since Q is an essential extension of the module X. Let $0 \neq x = (\alpha - \beta)(m) \in X \cap (\alpha - \beta)(M)$, where $m \in M$. Since $\alpha(m) = (\alpha - \beta)(m) + \beta(m) = x + \beta(m) \in M$, we have $m \in X$. Therefore, $(\alpha - \beta)(m) = 0$ and $x = 0$. This is a contradiction.

b. The assertion follows from **a.**

c. Since $\alpha = \alpha^2$, we have that $f = f^2$ and $f^2(x) = f(x) \in X$ for any element $x \in X$. Therefore, f is an idempotent endomorphism of the module X. By assumption, f can be extended to an endomorphism g of the module M. By **b** $\alpha(M) \subseteq M$. □

6.1.5 Automorphism-invariant, strongly automorphism-extendable and automorphism-extendable modules. We recall some notions from the introduction.

A module M is said to be *automorphism-invariant* if M is a characteristic submodule of its injective hull.

A module M is said to be *strongly automorphism-extendable* if for any submodule X in M, every automorphism of the module X can be extended to an automorphism of the module M.

A module M is said to be *automorphism-extendable* (resp., *endomorphism-extendable*) if for any submodule X in M, every automorphism of the module X can be extended to an endomorphism of the module M.

A module M is said to be *strongly endomorphism-extendable* if for any submodule X in M, every homomorphism $X \to M$, which maps into itself from some essential submodule of X, can be extended to a homomorphism $M \to M$.

a. Every characteristic submodule of any automorphism-invariant (resp., strongly automorphism-extendable) of the module is an automorphism-invariant (resp., strongly automorphism-extendable) module.

b. It follows from 6.1.3 that every automorphism-invariant module is strongly automorphism-extendable.

The converse assertion is not true in general case, since $\mathbb{Z}$ is a strongly automorphism-extendable $\mathbb{Z}$-module which is not automorphism-invariant.
Indeed, the additive group of rational numbers $\mathbb{Q}$ is the injective hull of the module $\mathbb{Z}_{\mathbb{Z}}$. In addition, $\mathbb{Z}$ is not an automorphism-invariant module, since $\alpha(\mathbb{Z}) \not\subseteq \mathbb{Z}$, where $\alpha\colon q \to q/2$ is an automorphism $\mathbb{Z}$-module $\mathbb{Q}$. Nevertheless, it is directly verified that any nonidentity automorphism α of an arbitrary nonzero submodule X of the module $\mathbb{Z}_{\mathbb{Z}}$ is a multiplication by -1; therefore, α can be extended to an automorphism of the module $\mathbb{Z}_{\mathbb{Z}}$.

c. It is clear that every strongly automorphism-extendable (resp., strongly endomorphism-extendable) module is an automorphism-extendable (resp., endomorphism-extendable). The author does not know an example of an automorphism-extendable module, which is not a strongly automorphism-extendable (resp., strongly endomorphism-extendable).

d. It is clear that every fully invariant submodule of any quasi-injective (resp., endomorphism-extendable, strongly endomorphism-extendable) of the module is a quasi-injective (resp., endomorphism-extendable, strongly endomorphism-extendable) module.

e. It is directly verified that $\mathbb{Z}$ is a strongly endomorphism-extendable nonquasi-injective $\mathbb{Z}$-module.

6.1.6. Let M be a module, Q be its injective hull, X be a submodule in M, and let Y be an arbitrary $\cap$-complement for the module X in the module M. Then $X' = X \oplus Y$ is an essential submodule in M.

a. For any submodule X' in M, every automorphism (resp., endomorphism) f of the module X can be extended to an automorphism (resp., the endomorphism) f' of an essential submodule X' of the module M with the use of the relation $f'(x+y) = f(x)+y$, where $x \in X$ and $y \in Y$. In its turn, an automorphism (resp., endomorphism) f' of the module X' can be extended to an automorphism (resp., the endomorphism) α of the injective of the module Q.

b. With the use of the relation $g'(x+y) = g(x)$ ($x \in X$ and $y \in Y$), any homomorphism $g\colon X \to M$ with essential in X kernel K can be extended to a homomorphism $g'\colon X' \to M$ with essential in M kernel $K' = K \oplus Y$. In its turn, the homomorphism g' can be extended to an endomorphism h of the injective module Q with essential in Q kernel and $1_Q - h$ is an automorphism of the module Q which coincides with the identity automorphism on the essential submodule K' of the module M.

c. The module M is strongly automorphism-extendable if and only if for any essential submodule X in M, every automorphism of the module X can be extended to an automorphism of the module M.
In addition, the module M is automorphism-extendable (resp., endomorphism-extendable) if and only if for any essential submodule X in M, every automorphism

(resp., endomorphism) of the module X can be extended to an endomorphism of the module M.

Proof. Since X' is an essential submodule in M and M is an essential submodule in Q, we have that X' is an essential submodule in Q.

a. It is directly verified that f' is an automorphism (resp., endomorphism) of the essential submodule X' of the module Q. Since the module Q is injective, an automorphism (resp., endomorphism) f' of the module X' can be extended to an endomorphism α of the module Q. We assume that f' is an automorphism of the module X'. Since X' is an essential submodule in Q and $X \cap \operatorname{Ker} \alpha = 0$, we have that α is a monomorphism, the module $\alpha(Q)$ is injective. Therefore, $\alpha(Q)$ is a direct summand of Q. In addition, $X' = f'(X') = \alpha(X'))$, whence the module $\alpha(Q)$ contains the essential submodule X' of the module Q. Therefore, $\alpha(Q) = Q$ and α is an automorphism.

Let f' be an endomorphism (resp., automorphism) of the module X', Y be a submodule in M which is maximal among submodules of M that have the zero intersection with X'. Then $X = X' \oplus Y$ is an essential submodule in M. Now we can define an endomorphism (resp., automorphism) f of the module X such that $f(x' + y) = f(x') + y$ for any $x' \in X'$ and $y \in Y$.

b. It is directly verified that $g' : X' \to M$ is a homomorphism with an essential M kernel $K' = K \oplus Y$. Since the module Q is injective, the homomorphism g' can be extended to an endomorphism h of the module Q. Since the module $\operatorname{Ker} h$ contains the essential submodule K' of the module M, we have that $\operatorname{Ker} h$ is an essential submodule in Q. Then the restriction of the endomorphism $1_Q - h$ of the injective module Q to the module $\operatorname{Ker} h$ is the identity automorphism of the essential submodule $\operatorname{Ker} h$ of the module Q. Then $1_Q - h$ is an essential injective submodule of the module Q. Therefore, $1_Q - h$ is an automorphism of the module Q, which coincides with the identity automorphism on the essential submodule K' of the module M.

c. The assertion follows from **a.** □

6.1.7 Semi-Artinian modules and semiprimary rings. A module M is said to be *semi-Artinian* if each of its nonzero factor modules has a simple submodule.

A semilocal ring with nilpotent Jacobson radical is called a *semiprimary* ring. In particular, every right or left Artinian ring is semiprimary.

The given below assertions are well known and are directly verified.

a. Every right module over a semi-Artinian right ring is a semi-Artinian module, and all semiprimary rings are semi-Artinian.
In particular, every module over a semiprimary ring is semi-Artinian.

b. If M is a module such that all its cyclic submodules are Artinian, then M is a semi-Artinian module.
In particular, every Artinian module is semi-Artinian.

c. Every Abelian torsion group is a semi-Artinian $\mathbb{Z}$-module, and any direct sum of infinitely many nonzero Abelian torsion groups is a non-Artinian semi-Artinian $\mathbb{Z}$-module.

d. There exist semi-Artinian rings, which are not semiprimary.
Indeed, let F be a field and A the ring of all sequences elements of F, which are stabilized at finite number depending on the sequence. Then A is a commutative semi-Artinian ring which is not a semiprimary ring.

6.1.8 Theorem ([184]). For a semi-Artinian module M, the following conditions are equivalent.

1) M is an automorphism-invariant module.

2) M is a strongly automorphism-extendable module.

3) M is an automorphism-extendable module.

Proof. The implication 1) $\Rightarrow$ 3) follows from 6.1.5(b).

3) $\Rightarrow$ 2). Let X_1 be a submodule in M and f_1 its automorphism. We have to prove that f_1 can be extended to an automorphism of the module M. By 6.1.6(b), we can assume that X_1 is an essential submodule in M.

We denote by W the set of all pairs (X', f') such that X' is a submodule in M, $X_1 \subseteq X'$, f' is an automorphism of the module X' and f' coincides with f_1 on X_1. We define a partial order on W such that $(X', f') \leq (X'', f'')$ in only if $X' \subseteq X''$ and f'' coincides with f' on X'. It is directly verified that the union of any ascending chain of pairs from W belongs to W. By the Zorn lemma, there exists a maximal pair (X, f). Then $X = X_2$ for any pair $(X_2, f_2) \in W$ such that $X \subseteq X_2$ and f_2 coincides with f on X.

If $X = M$, then f is an automorphism of the module M, which is required.

We assume that $X \neq M$. Then the nonzero semi-Artinian module M/X is an essential extension of its nonzero socle Y/X, where X is a proper submodule of the module $Y \subseteq M$. Since M is an automorphism-extendable module, the automorphisms f and f^{-1} of the module X can be extended to endomorphisms g and h of the module M, respectively. Then $(1 - gh)(X) = 0 = (1 - hg)(X)$. Since $X \cap \operatorname{Ker} g = X \cap \operatorname{Ker} f = 0$ and X is an essential submodule in M, we have that g is a monomorphism. In addition, the restriction of g to X is an automorphism of the module X. Therefore, g induces the monomorphism $\bar{(g)}\colon M/X \to M/X$. Since the socle Y/X of the module M/X is a fully invariant submodule in M/X, we have $\bar{(g)}(Y/X) \subseteq Y/X$. Therefore, $g(Y) \subseteq Y$ and $X = g(X) \subsetneq g(Y)$. Similarly, we obtain that $h(Y) \subseteq Y$ and $X = h(X) \subsetneq h(Y)$.

Since M is a semi-Artinian module, M is an essential extension of its nonzero socle S. Then $S \subseteq X$, since X is an essential submodule in M and S is a semisimple submodule in M. In addition, $(1 - gh)(X) = (1 - ff^{-1})(X) = 0$ and Y/X is a semisimple module. Therefore, $(1 - gh)(Y) \subseteq S \subseteq X$. Then

$$Y \subseteq (1 - gh)(Y) + gh(Y) \subseteq X + g(Y) = g(Y)\,.$$

Therefore, the restriction of g_Y of the monomorphism g to Y is an automorphism of the module Y and g_Y is an extension of the automorphism f of the module X. This contradicts the choice of X.

2) ⇒ 1). Let Q be the injective hull of the module M, u be an automorphism of the module Q, and let S be the socle of the module Q. We have to prove that $u(M) \subseteq M$.

Since M is an essential semi-Artinian submodule in Q, we have that S is contained in M; it is an essential submodule in M. Since S is the socle of the module Q, we have $u(S) = S = u^{-1}(S)$. We denote by X the sum of all submodules X' in M such that $u(X') = X' = u^{-1}(X')$. Then $u(X) = X = u^{-1}(X)$, $S \subseteq X$, X is an essential submodule in M.

If $X = M$, then $u(M) = M$, which is required.

We assume that $X \neq M$. Then the nonzero semi-Artinian module M/X is an essential extension of its nonzero socle Y/X, where $X \subsetneq Y \subseteq M$. Since $u(X) = X = u^{-1}(X)$ and M is a strongly automorphism-extendable module, there exist automorphisms f, g of the module M such that $(u - f)(X) = 0$ and $(u^{-1} - g)(X) = 0$. In addition, Y/X is a semisimple module. Therefore, $(u - f)(Y) \subseteq S \subseteq X$ and $(u^{-1} - g)(Y) \subseteq S \subseteq X$. Since f, g are automorphisms of the module M and $f(X) = X = g(X)$, we have that f and g induce the automorphisms $\overline{f}$ and $\overline{g}$ of the module M/X with nonzero socle Y/X. Since Y/X is the socle of the module M/X, we have $\overline{f}(Y/X) = Y/X = \overline{g}(Y/X)$ and $f(X) = X = g(X)$. Therefore, $f(Y) = Y = g(Y)$. In addition, $(u - f)(Y) \subseteq Y$ and $(u^{-1} - g)(Y) \subseteq Y$. Then $u(Y) \subseteq (u - f)(Y) + f(Y) \subseteq Y$ and $u^{-1}(Y) \subseteq (u^{-1} - g)(Y) + g(Y) \subseteq Y$. Therefore, $u(Y) = Y = u^{-1}(Y)$ and Y properly contains X; this contradicts to the choice of X. □

6.1.9 Corollary ([174, 181]). Let A be a ring and M an A-module. If M is an Artinian module or A is a semiprimary ring, then the following conditions are equivalent.

1) M is an automorphism-invariant module.

2) M is a strongly automorphism-extendable module.

3) M is an automorphism-extendable module.

6.1.10 Theorem. For a semi-Artinian module M, the following conditions are equivalent.

1) M is a quasi-injective module.

2) M is a strongly endomorphism-extendable module.

3) M is an endomorphism-extendable module.

Proof. The implications 1) ⇒ 2) and 2) ⇒ 3) are always true.

3) ⇒ 1). Let Q be the injective hull of the module M, S be the socle of the module Q, and let f be an endomorphism of the module Q. It is sufficient to prove that $f(M) \subseteq M$. We remark that $f(S) \subseteq S$. In addition, $S \subseteq M$, since M is an essential submodule in Q.

We denote by X the sum of all submodules X' in M such that $f(X') \subseteq X'$. Then $f(X) \subseteq X$, $S \subseteq X$ and X is an essential submodule in M.

If $X = M$, then $f(M) \subseteq M$, which is required.

We assume that $X \neq M$. Then the nonzero semi-Artinian module M/X is an essential extension of its nonzero socle Y/X, where $X \subsetneq Y \subseteq M$. Since $f(X) \subseteq X$ and M is an endomorphism-extendable module, there exists an endomorphism g of the module M such that $(f - g)(X) = 0$. In addition, Y/X is a semisimple module. Therefore, $(f - g)(Y) \subseteq S \subseteq X$. Since $f(X) \subseteq X)$, we have that f induces the endomorphism $\overline{f}$ of the module M/X with nonzero socle Y/X. Since Y/X is the socle of the module M/X, we have $\overline{f}(Y/X) \subseteq Y/X$ and $f(X) \subseteq X$. Therefore, $f(Y) \subseteq Y$. In addition, Y properly contains X; this contradicts the choice of X. □

6.1.11. Let A be a ring, M be an automorphism-extendable right A-module, and let X, Y be two submodules in M with $X \cap Y = 0$.

a. If $f\colon Y \to X$ is a homomorphism, then there exists an endomorphism g of the module M that coincides with $f\colon Y \to X$ on Y.

b. If $M = X \oplus Y$, then the module X is injective with respect to Y.

Proof. **a.** We define an endomorphism α of the module $X \oplus Y$ by the relation $\alpha(x+y) = x + f(y) + y$ for all $x \in X$ and $y \in Y$. We assume that

$$0 = \alpha(x + y) = x + f(y) + y\,, \quad x \in X,\ y \in Y\,.$$

Then α is a monomorphism, since

$$y = -x - f(y) \in X \cap Y = 0\,, \quad f(y) = 0\,,$$
$$x = x + f(y) + y = \alpha(x + y) = 0\,.$$

In addition, for any $x \in X$ and $y \in Y$, we have

$$x + y = (x - f(y)) + (f(y) + y) = \alpha(x - f(y)) + \alpha(y) \in \alpha(X \oplus Y)\,.$$

Therefore, α is an automorphism of the module $X \oplus Y$. In addition, the endomorphism $\alpha - 1_{X\oplus Y}$ of the module $X \oplus Y$ coincides with the homomorphism $f\colon Y \to X$ on Y. Since M is an automorphism-extendable module, an automorphism α of the module $X \oplus Y$ can be extended to an endomorphism β of the module M. We denote by g the endomorphism $\beta - 1_M$ of the module M. Then g coincides with f on Y.

b. Let Y_1 be a submodule in Y and f_1 a homomorphism from Y_1 into X. By **a**, there exists an endomorphism g of the module M, which coincides with $f_1\colon Y_1 \to X$ on Y_1. Let π be the projection of the module $M = X \oplus Y$ onto X with kernel Y and let $u\colon Y \to M$ be a natural embedding. We denote by f the homomorphism $\pi g u$ from Y into X. Then f coincides with f_1 on Y_1. Therefore, the module X is injective with respect to Y. □

6.1.12. If M is an automorphism-invariant uniform module, then M is a quasi-injective module.

Proof. Let Q be the injective hull of the module M and $f \in \operatorname{End} Q$. It is sufficient to prove that $f(m) \in M$ for any $m \in M$. It follows from the definition of an automorphism-invariant module that the required inclusion is true if f is an automorphism. We assume that f is not an automorphism. Then $1_Q - f$ is an automorphism, since ring $\operatorname{End} Q$ is local by 6.1.2(c). Therefore, $(1_Q - f)(m) \in M$. Then $f(m) = m - (1_Q - f)(m) \in M$. □

6.1.13. Let M be an automorphism-extendable module and $M = \oplus_{i \in I} M_i$.

a. If M_i is a quasi-injective module for any $i \in I$, then M is a quasi-injective module.

b. If M_i is an automorphism-invariant uniform module for any $i \in I$, then M is a quasi-injective module.

Proof. **a.** Since M is an automorphism-extendable module and $M = M_i \oplus_{j \neq i} M_j$ for any $i \in I$, it follows from 6.1.11(b) that for any $i \in I$, the module $\oplus_{j \neq i} M_j$ is M_i-injective. By 6.1.1(k), M is a quasi-injective module.

b. We fix $i \in I$. By 6.1.12, M_i is a quasi-injective module. By **a**, M is a quasi-injective module. □

6.1.14. Let A be an Artinian serial ring. It is well known, e.g., see [63, 25.4.2] or [194, 55.16] that every A-module M is the direct sum of uniserial modules of finite (composition) length. In particular, M is a direct sum of uniform modules and if M is an indecomposable module, then M is a uniserial Artinian and Noetherian module.

6.1.15 Theorem. Let A be a serial Artinian ring and let M be an A-module. The following conditions are equivalent.

1) M is an automorphism-extendable module.

2) M is a strongly automorphism-extendable module.

3) M is an automorphism-invariant module.

4) M is an endomorphism-extendable module.

5) M is a strongly endomorphism-extendable module.

6) M is a quasi-injective module.

Proof. The implications $6) \Rightarrow 3) \Rightarrow 2) \Rightarrow 1)$ and $6) \Rightarrow 5) \Rightarrow 4) \Rightarrow 1)$ are true for all modules over any ring.

$1) \Rightarrow 6)$. By 6.1.14, $M = \sum_{i \in I} M_i$, where all M_i are uniserial modules of finite length. We fix $i \in I$. Since A is an Artinian ring, M_i is a semi-Artinian module. By 6.1.13(c), M is a quasi-injective module. □

6.1.16 Remark. If A is a ring and every right A-module is the direct sum of uniform modules, then every automorphism-extendable right A-module M is a quasi-injective module.

Indeed, A is an Artinian ring; e.g., see [59, Theorem 1]. Therefore, every A-module is semi-Artinian. By 6.1.13(c), M is a quasi-injective module.

6.2 Singular and nonsingular modules

6.2.1 The singular submodule Sing M. The ideals sg A_A, sg$_A$ A and sg M. If M is a right (resp., left) module over the ring A, then we denote by Sing M the set of all elements $m \in M$ such that $r(m)$ (resp., ($\ell(m)$)) is an essential right (resp., left) ideal of the ring A.

A module M is said to be *singular* if $\operatorname{Sing} M = M$. A module M is said to be *nonsingular* if $\operatorname{Sing} M = 0$.

We denote by $G(M)$ or $\operatorname{Sing}_2 M$ the intersection of all submodules X of the module M such that the factor module M/X is nonsingular. The submodule $G(M)$ is called the *Goldie radical* or the *second singular submodule* of the module M. A module M is said to be *Goldie-radical* if $G(M) = M$. The relation $G(M) = 0$ is equivalent to the relation $\operatorname{Sing} M = 0$.

In the assertion **b** below, it is proven that Sing M is a fully invariant submodule in M; it is called *(the largest) singular submodule* of the module M. Therefore, for any ring A, the sets $\operatorname{Sing} A_A$ and $\operatorname{Sing} {}_AA$ are ideals that are called the *right singular ideal* and the *left singular ideal*, respectively.

The set of all endomorphisms of the module M, whose kernels are essential submodules in M, is denoted by sg M. In the assertion **b** below, it is proven that sg M is an ideal of the ring End M.

Let A be a ring and M a right A-module.

a. If N is an essential submodule of the module M, then the module M/N is singular.

b. If N is an essential submodule of the module M, then for every homomorphism $f\colon X \to M$, the submodule $f^{-1}(N)$ is essential in X.

c. If L is a right A-module and T is the subset of the homomorphism group $\operatorname{Hom}(L_A, M_A)$ consisting of all homomorphisms t such that Ker t is an essential submodule in L, then T is a subbimodule of the End M-End L-bimodule $\operatorname{Hom}(L, M)$.

Consequently, sg M is an ideal of the ring End M, $\operatorname{Sing} A_A = \operatorname{sg} A_A$, $\operatorname{Sing} {}_AA = \operatorname{sg} {}_AA$ and $\sum_{t\in T} t(L)$ is a fully invariant submodule in M_A and this submodule is contained in Sing M. In particular, if M is a nonsingular module, then sg $M = 0$.

d. Sing M is a submodule in M that contains all singular submodules in M, and $g(\text{Sing } M) \subseteq \text{Sing } Y$ for every module homomorphism $g\colon M \to M'$.
In particular, Sing M is a fully invariant singular submodule in M and $\text{Sing } X = G(X) = 0$ for any submodule X in M such that $X \cap \text{Sing } M = 0$.

e. Let the ring A be right uniform. Then its right singular ideal $\text{Sing } A_A$ is a completely prime ideal and coincides with the set N_ℓ of all left zero divisors of the ring A. In addition, if A is right nonsingular, then A is a right uniform domain.

Proof. **a.** Let $h\colon M \to M/N$ be the natural epimorphism, $m \in M$ and $B = r(h(m))$. We assume that the right ideal B of the ring A is not essential. Then $B \cap C = 0$ for some nonzero right ideal C of the ring A. Since $C \not\subseteq r(h(m))$, we have $h(mC) \neq h(0)$ and $mC \neq h(0)$. Since N is an essential submodule of the module M, we have $N \cap mC \neq 0$. Therefore, there exists a nonzero right ideal D such that $D \subseteq C$ and $0 \neq mD \subseteq N$. Then $D \subseteq r(h(m)) = B$. Then $D \subseteq B \cap C = 0$. This is a contradiction.

b. We assume that there exists a nonzero submodule Y in X such that $Y \cap f^{-1}(N) = 0$. Since $\text{Ker} f \subseteq f^{-1}(N)$, we have $Y \cap \text{Ker} f = 0$. Therefore, $f(Y) \neq 0$. Then $N \cap f(Y) \neq 0$. Since $Y \cap \text{Ker} f = 0$, there exists a nonzero submodule Y_1 in Y such that $f(Y_1) = N \cap f(Y)$. Then $0 \neq Y_1 \subseteq Y \cap f^{-1}(N) = 0$. This is a contradiction.

c. We assume that $t, u \in T$, $f \in \text{End}(L)$ and $g \in \text{End } M$. It is sufficient to prove that $t + u \in T$, $gt \in T$, and $tf \in T$. The inclusion $t + u \in T$ follows from the property that $\text{Ker } t + u)$ contains the essential submodule $\text{Ker } t \cap \text{Ker } u$. Since $\text{Ker}(gt)$ contains the essential submodule $\text{Ker } t$ in M, we have $gt \in T$. Since L is an essential extension of $\text{Ker } t$, it follows from **b** that L is an essential extension of the module $f^{-1}(\text{Ker } t)$. Since $\text{Ker}(tf) \supseteq f^{-1}(\text{Ker } t)$, we have $tf \in T$, which is required.

d. We define a mapping $f\colon M \to \text{Hom}(A_A, M)$ such that $f(m)(a) = ma$ for all $m \in M$ and $a \in A$. It is directly verified that f is an End M-A-bimodule isomorphism. Therefore, **d** follows from **c**.

e. Since the ring A is right uniform, the ideal $\text{Sing } A_A$ coincides with N_ℓ; therefore, N_ℓ is a completely prime ideal. □

6.2.2 Closed submodules and closures. Let M be a module and X a submodule in M. X is called a *closed* submodule in M if $X = X'$ for every submodule X' in M that is an essential extension of the module X.

A closed submodule $\overline{X}$ of the module M is called the *closure* of the module X in M if $\overline{X}$ is an essential extension of the module X.

It is well known that X has at least one closure in M and there exists at least one closed submodule Y in M which is called $\cap$-*complement* to X in M such that $X \cap Y = 0$ and $X \oplus Y$ is an essential submodule in M. Under these conditions, Y also is a $\cap$-complement for $\overline{X}$ in M.

a. Let $X = \oplus_{i \in I} X_i$, $\bar{X}_i$ be a closure of the module X_i in M, $i \in I$, and let $X^* = \sum_{i \in I} \bar{X}_i$. Then $X^* = \oplus_{i \in I} \bar{X}_i$.

b. Let $X = X_1 \oplus X_2 \oplus \cdots \oplus X_n$, $\bar{X}_i$ be a closure of the module X_i in M, $i = 1, 2, \ldots, n$, $X^* = X_1 \oplus X_2 \oplus \cdots \oplus X_n \subseteq M$, and let Q be the injective hull of the module M. Then there exists a direct decomposition $Q = Q_1 \oplus Q_2 \oplus \cdots \oplus Q_n \oplus P$ such that $\overline{X}_i = M \cap Q_i$, $i = 1, \ldots n$ and $Q_1 \oplus Q_2 \oplus \cdots \oplus Q_n$ is the injective hull of the module X. If X is an essential submodule in M, then $Q_1 \oplus Q_2 \oplus \cdots \oplus Q_n$ is the injective hull of the module M.

c. $G(M)$ is a fully invariant submodule of the module M, the right Goldie radical $G(A_A)$ and the left Goldie radical $G({}_AA)$ of the ring A are ideals of the ring A, and $f(G(M)) \subseteq G(M)$ for any homomorphism $f\colon G(M) \to M$.

d. $G(M)$ is a closed submodule in M; it is a closure of the singular submodule $\operatorname{Sing} M$ in M.

e. If the ring A is right nonsingular, then $G(M) = \operatorname{Sing} M$ and the module $M/\operatorname{Sing} M$ is nonsingular.

All assertions from 6.2.2 are well known and are directly verified; e.g., see [79] and [166].

6.2.3 Nonsingular and singular modules. Let A be a ring and M a right A-module.

a. If $f\colon X \to M$ is a module homomorphism such that $\operatorname{Ker} f$ is an essential submodule in M, then the module $f(X)$ is singular; therefore, it is contained in $\operatorname{Sing} M$. Consequently, if the module M is nonsingular, then the kernel of any nonzero module homomorphism $X \to M$ is not an essential submodule in X.

b. If X is a submodule in M and the module M/X is nonsingular, then X is a closed submodule in M.

c. If the module M is nonsingular, then its submodule X is closed in M if and only if the module M/X is nonsingular.

d. If the module M is nonsingular, then the kernel of any endomorphism f is a closed submodule in M.

e. All submodules, essential extensions and subdirect products of nonsingular modules are nonsingular.

f. If $\{X_i\}_{i \in I}$ is some set submodules in M and all modules M/X_i are nonsingular, then every module X_i is closed in M, the module $M/(\cap_{i \in I})X_i$ is nonsingular and $X = \cap_{i \in I} X_i$ is a closed submodule in M. In particular, the factor module $M/G(M)$ is nonsingular and $G(M)$ is a closed submodule in M.

g. If M is not a singular, then M contains an isomorphic copy of a nonzero right ideal of the ring A. In addition, if A is right finite-dimensional, then M contains an isomorphic copy of a nonzero uniform right ideal of the ring A.

Proof. **a.** The assertion follows from isomorphism $f(X) \cong X/\operatorname{Ker} f$ and the property that the module $X/\operatorname{Ker} f$ is singular by 6.2.1(a).

b. Let $Y \in \operatorname{Lat}(M)$ and let X be an essential submodule in Y. Since the module M/X is nonsingular and the module Y/X is singular by 6.2.1(a), we have that $Y/X = 0$, $X = Y$; therefore, X is closed in M.

c. If M/X is nonsingular, then X is closed in M by **b**.

We assume that X is closed in M and $Y/X = \operatorname{Sing}(M/X)$. If $Y = X$, then the assertion has been proven. We assume that Y properly contains the closed submodule X in M. Then $X \cap yA = 0$ for some nonzero $y \in Y$. Since $y+X \in \operatorname{Sing}(M/X)$, we have that $(y^{\cdot}.X)$ is an essential right ideal. Since $X \cap yA = 0$, we have that $r(y) = (y^{\cdot}.X)$ is an essential right ideal; this is a contradiction.

d. Let $X = \operatorname{Ker} f$. Since $M/X \cong f(M) \subseteq M$, we have that M/X is nonsingular and X is a closed submodule in M by **b**.

e. The assertion is directly verified.

f. Since M/X is the subdirect product of nonsingular modules M/X_i, it follows from **e** that the module M/X is nonsingular. By **b**, X is a closed submodule in M. Now the remaining assertions are directly verified.

g. By assumption, there exists a nonzero element $m \in M$ such that the right ideal $r(m)$ in A is not an essential. Let $h\colon A_A \to mA$ be the natural epimorphism and let B be a nonzero right ideal with $B \cap r(m) = 0$. Since $B \cap \operatorname{Ker} h = 0$, we have that B_A is isomorphic to the nonzero submodule $h(B)$ in M. If A is right finite-dimensional, then B contains a nonzero uniform right ideal C, whence M contains $h(C)$. □

6.2.4 ([179]). Let A be a ring and M an automorphism-extendable right A-module.

a. If X, Y are two submodules in M with $X \cap Y = 0$ and the module M/X is nonsingular, then the module X is injective with respect to Y.

b. If A is a right nonsingular ring and Y is any nonsingular submodule in M, then the module $\operatorname{Sing} M$ is injective with respect to Y.

c. If the module M is nonsingular and X, Y are two closed submodules in M with $X \cap Y = 0$, then the module X is injective with respect to Y and the module Y is injective with respect to X.

Proof. **a.** Let Y_1 be a submodule of the module Y and $f_1\colon Y_1 \to X$ a homomorphism. We have to prove that f_1 can be extended to a homomorphism $f\colon Y \to X$. Without loss of generality, we can assume that Y_1 is an essential submodule in Y. (Indeed, by the Zorn lemma, there exists a submodule Z in Y such that $Y_1 \cap Z = 0$ and $Y_1 \oplus Z$ is an essential submodule in Y. We set $Y_2 = Y_1 \oplus Z$. The homomorphism $f_1\colon Y_1 \to X$ can be extended to a homomorphism $f_2\colon Y_2 \to X$ with the use of the relation $f_2(y_1 + z) = f_1(y_1)$.)

We define an endomorphism α of the module $X \oplus Y_1$ by the relation $\alpha(x + y_1) = x + f(y_1) + y_1$ for all $x \in X$ and $y_1 \in Y_1$. We assume that

$$0 = \alpha(x + y_1) = x + f(y_1) + y_1 , \quad x \in X,\ y_1 \in Y_1 ,$$
$$y_1 = -x - f(y_1) \in X \cap Y_1 = 0 , \quad f(y_1) = 0 ,$$
$$x = x + f(y_1) + y_1 = \alpha(x + y_1) = 0 .$$

In addition, for any $x \in X$ and $y_1 \in Y_1$, we have

$$\begin{aligned} x + y_1 &= (x - f(y_1)) + (f(y_1) + y_1) \\ &= \alpha(x - f(y_1)) + \alpha(y_1) \in \alpha(X \oplus Y_1) . \end{aligned}$$

Therefore, α is an automorphism of the module $X \oplus Y_1$. It follows from the construction of α that the endomorphism $\alpha - 1$ of the module $X \oplus Y_1$ coincides with the homomorphism $f_1 \colon Y_1 \to X$ on the module Y_1. Since M is an automorphism-extendable module, the automorphism α of the module $X \oplus Y_1$ can be extended to an endomorphism β of the module M. We denote by g the endomorphism $\beta - 1$ of the module M. Then g coincides with f_1 on Y_1.

We prove that $g(y) \in X$ for any element y of the module Y. Let $h \colon M \to M/X$ be the natural epimorphism. Since Y_1 is an essential submodule in Y, it follows from 6.2.1(a) that $yB \subseteq Y_1$ for some essential right ideal B of the ring A. Then

$$g(y)B = g(yB) \subseteq g(Y_1) = f_1(Y_1)) \subseteq X ,$$
$$h(g(y))B = h(g(y)B) \subseteq h(X) = 0 .$$

Therefore, $h(g(y)) \in \mathrm{Sing}(M/X) = 0$ and $g(y) \in \mathrm{Ker}\, h = X$.

Since $g(Y) \subseteq X$, we have that g induces the homomorphism $f \colon Y \to X$. Therefore, the module X is injective with respect to Y.

b. We set $X = \mathrm{Sing}\, M$. Since A is a right nonsingular ring, it follows from 6.2.2(e) that the module M/X is nonsingular. Since the module X is singular and the module Y is nonsingular, it follows from **a** that the module X is injective with respect to Y.

c. Since X, Y are closed submodules of the nonsingular module M, it follows from 6.2.3(c) that the modules M/X and M/Y are nonsingular. By **a**, X is injective with respect to Y and the module Y is injective with respect to X. □

6.2.5. Let M be a module and let Q be its injective hull. The following conditions are equivalent.

1) M is an automorphism-extendable module and for any essential submodule Y in M, every homomorphism $Y \to M$ with essential in Y kernel can be extended to an endomorphism of the module M with essential in M kernel.

2) M is a strongly automorphism-extendable module and for each submodule Y in M, any homomorphism $Y \to M$ with essential in Y kernel can be extended to an endomorphism of the module M with essential in M kernel.

3) $\alpha(M) \subseteq M$ for any automorphism α of the module Q such that $\alpha(X) = X$ for some essential submodule X of the module M.

4) $\alpha(M) = M$ for any automorphism α of the module Q such that $\alpha(X) = X$ for some essential submodule X of the module M.

Proof. The implications 4) $\Rightarrow$ 3) and 2) $\Rightarrow$ 1) are directly verified.

1) $\Rightarrow$ 3). Let α be an automorphism of the module Q and $\alpha(X) = X$ for some essential submodule X of the module M. We denote by Y the submodule

$$\alpha^{-1}(M \cap \alpha(M)) = \{y \in M \mid \alpha(y) \in M\}$$

of the module M. Then $\alpha(Y) \subseteq M$, $X \subseteq Y$, and Y is an essential submodule in M. In addition, α induces the automorphism φ_1 of the module X. Since M is an automorphism-extendable module, φ_1 can be extended to an endomorphism φ_2 of the module M. Since the module Q is injective, φ_2 can be extended to an endomorphism φ of the module Q. We denote by g the restriction of the homomorphism $\alpha - \varphi$ to the module Y. Since $\varphi(Y) \subseteq M$, $\alpha(Y) \subseteq M$ and $g(X) = 0$, we have that g is a homomorphism from Y into M with an essential Y kernel. By assumption, g can be extended to an endomorphism g_1 of the module M. Since the module Q is injective, g_1 can be extended to an endomorphism β of the module Q. Then $(\alpha - \varphi - \beta)(Y) = (g - \beta)(Y) = 0$. We denote by Z the submodule $\{z \in M \mid (\alpha - \varphi - \beta)(z) \in M\}$ of the module M. Then Z is the complete pre-image in M of the module $M \cap (\alpha - \varphi - \beta)(M)$ under the action of the homomorphism $\alpha - \varphi - \beta$. In addition, $Y \subseteq Z$ and

$$\alpha(Z) \subseteq (\alpha - \varphi - \beta)(Z) + (\varphi + \beta)(Z) \subseteq M\,.$$

Therefore, $Y \subseteq Z \subseteq Y$ and $Z = Y$. Then

$$(\alpha - \varphi - \beta)(Z) = (\alpha - \varphi - \beta)(Y) = 0\,.$$

If $(\alpha - \varphi - \beta)(M) = 0$, then $\alpha(M) = (\varphi - \beta)(M) \subseteq M$, which is required.

We assume that $(\alpha-\varphi-\beta)(M) \neq 0$. Since M is an essential submodule in Q, we have that $M \cap (\alpha - \varphi - \beta)(M)$ is an essential submodule of the nonzero module $(\alpha - \varphi - \beta)(M)$. Since Z is the complete pre-image in M of the nonzero module $M \cap (\alpha - \varphi - \beta)(M)$ under the action of the homomorphism $\alpha - \varphi - \beta$, we have $(\alpha - \varphi - \beta)(Z) \neq 0$. This is a contradiction.

3) $\Rightarrow$ 4). Let X be an essential submodule of the module M and let α be an automorphism of the module Q with $\alpha(X) = X$. It follows from 3) that $\alpha(M) \subseteq M$ and $\alpha^{-1}(M) \subseteq M$. Then $\alpha(M) = M$.

4) $\Rightarrow$ 2). Let X be a submodule of the module M and φ an automorphism of the module X. By 6.1.6(a) and the property that M is an essential submodule in Q, we can assume that X is an essential submodule in Q. By 6.1.6(a), the automorphism φ of the module

X can be extended to an automorphism α of the injective of the module Q. By assumption, $\alpha(M) = M$. Therefore, φ can be extended to an automorphism of the module M.

Let $h_1: Y \to M$ be a homomorphism with an essential Y kernel K_1. By 6.1.6(b), we can assume that Y, K_1 are essential submodules in Q. The homomorphism h_1 can be extended to an endomorphism h of the injective of the module Q and h has an essential kernel K. We denote by α the endomorphism $1_Q - h$ of the module Q. Then the restriction of the endomorphism $1_Q - h$ to K is the identity automorphism of the essential submodule K of the injective module Q. Therefore, $\mathrm{Ker}(1_Q - h) = 0$ and $(1_Q - h)(Q)$ is an injective essential submodule in Q. Therefore, $1_Q - h$ is an automorphism of the module Q and $(1_Q - h)(K) = K$. By 4), $(1_Q - h)(M) = M$. Therefore, $(1_Q - h)|_M$ is an automorphism of the module M. Then $1_M - (1_Q - h)|_M$ is an endomorphism of the module M with essential in M kernel and $1_M - (1_Q - h)|_M$ coincides with h_1 on Y. □

6.2.6 Theorem ([176]). Let M be a module, Q be the injective hull of the module M, and let $M = T \oplus U$, where T is an injective module and U is a nonsingular module. The following conditions are equivalent.

1) M is an automorphism-extendable module.

2) M is a strongly automorphism-extendable module.

3) $\alpha(M) \subseteq M$ for any automorphism α of the module Q such that $\alpha(X) = X$ for some essential submodule X of the module M.

Proof. The implication 3) ⇒ 2) follows from 6.2.5.

The implication 2) ⇒ 1) is obvious.

1) ⇒ 3). Let Y be an essential submodule in M, $h: Y \to M$ be a homomorphism with essential in Y kernel, and let $\pi: M = T \oplus U \to U$ be the projection with kernel T. Then the module $\pi h(Y)$ is singular and it is contained in the nonsingular module U. Therefore, $\pi h(Y) = 0$. Therefore, $h(Y) \subseteq T$ and h is a homomorphism from the module Y into the module T. Since the module T is injective, h can be extended to a homomorphism $M \to T \subseteq M$. This homomorphism is the required endomorphism of the module M, which extends h. □

6.2.7 Corollary ([176]). Let M be a nonsingular module and let Q be its injective hull. The following conditions are equivalent.

1) M is an automorphism-extendable module.

2) M is a strongly automorphism-extendable module.

3) $\alpha(M) \subseteq M$ for any automorphism α of the module Q such that $\alpha(X) = X$ for some essential submodule X of the module M.

Corollary 6.2.7 follows from Theorem 6.2.6 and the property that the nonzero module is injective.

6.2.8 Theorem ([176]). Let $M = T \oplus U$, where T is an injective module, U is a nonsingular module, and $\mathrm{Hom}(T', U) = 0$ for any submodule T' of the module T. The following conditions are equivalent.

1) M is an automorphism-extendable module.

2) M is a strongly automorphism-extendable module.

3) U is an automorphism-extendable module.

4) U is a strongly automorphism-extendable module.

Proof. The equivalence 1) $\Leftrightarrow$ 2) follows from Theorem 6.2.6.

The implication 1) $\Rightarrow$ 3) is directly verified.

The equivalence 3) $\Leftrightarrow$ 4) follows from Corollary 6.2.7.

3) $\Rightarrow$ 1). Let Q be the injective hull of the module M, X be an essential submodule of the module M, and let α be an automorphism of the module Q such that $\alpha(X) = X$. By Theorem 6.2.6, it is sufficient to prove that $\alpha(M) \subseteq M$. For the injective hull Q of the module $M = T \oplus U$, there exists a direct decomposition $Q = T \oplus U_1$, where U_1 is the injective hull of the nonsingular module U. Since α is an automorphism of the module $T \oplus U_1$ and $\mathrm{Hom}(T', U) = 0$ for any submodule T' of the module T, it is directly verified that $\alpha(T) = T$. Let $h\colon Q \to Q/T$ be the natural epimorphism. (We can assume that h is the projection of the module $Q = T \oplus U_1$ onto the module U_1 with kernel T.) Then α induces the automorphism α_1 of the injective hull $h(Q)$ of the module $h(U)$. Since $\alpha(X) = X$, we have $\alpha_1(h(X)) = h(X)$. By applying Corollary 6.2.7 to the automorphism-extendable nonsingular module $h(M) = h(U)$, we obtain that $\alpha_1(h(M)) \subseteq h(M)$. Then $\alpha(M) \subseteq M + T = M$. □

6.2.9 Corollary. Let $M = T \oplus U$, where T is an injective module, which is an essential extension of the singular module, and U is a nonsingular module. The following conditions are equivalent.

1) M is an automorphism-extendable module.

2) M is a strongly automorphism-extendable module.

3) U is an automorphism-extendable module.

4) U is a strongly automorphism-extendable module.

6.2.10 The completion of the proof of Theorems 6A, 6B and 6C. Theorems 6A and 6B follow from Theorems 6.1.8 and 6.1.15. Theorem 6C follows from Theorem 6.2.8.

6.2.11 Open question. Is it true that every automorphism-extendable module is strongly automorphism-extendable?

7 Modules over strongly prime and strongly semiprime rings

The main results of this section are Theorems 7A and 7B.

7A Theorem (Tuganbaev [179]). If A is a right strongly prime ring, then a right A-module M is automorphism-invariant if and only if either M is a singular automorphism-invariant module or M is an injective module.

7B Theorem (Tuganbaev [176]). If M is a right module over an invariant hereditary domain A, then the following conditions are equivalent.

1) M is an automorphism-extendable (strongly automorphism-extendable) module.
2) M is an endomorphism-extendable (strongly endomorphism-extendable) module.
3) Either M is a quasi-injective singular module or M is an injective module which is not singular, or $M = X \oplus Y$, where X is an injective singular module and the module Y is isomorphic to nonzero submodule in Q_A, where Q is the division ring of fractions of the domain A.

Remark. For the completion of the proof of Theorems 7A and 7B, see 7.2.15.

7.1 Modules over strongly semiprime rings

7.1.1 Annihilators and annihilator conditions. Let A be a ring and B a subset in A.

a. Let B be a subset in A. Then $r(\ell(r(B))) = r(B)$ and $\ell(r(\ell(B))) = \ell(X)$.

b. If A is a subring of some ring Q, then $r_A(B) = A \cap r_Q(B)$ and $\ell_A(B) = A \cap \ell_Q(B)$.

c. A is a ring with the maximum condition on left annihilators if and only if A is a ring with the minimum condition on right annihilators.

d. If A is a ring with the maximum condition on right annihilators, then its right singular ideal $\operatorname{Sing} A_A$ is nilpotent, $\ell(\operatorname{Sing}({}_AA))$ is an essential left ideal, every subset $B \subseteq A$ contains a finite subset $B' = \{b_1, \dots, b_n\}$ such that $\ell(B) = \ell(B') = \cap_{i=1}^{n} \ell(b_i)$, and eAe is a ring with the maximum condition on right annihilators for any nonzero idempotent $e \in A$.

Proof. **a, b.** The assertions are well known and are directly verified.

c. Let A be a ring with the maximum condition on right annihilators and let $\{\ell(X_i)\}_{i=1}^{\infty}$ be a set of left annihilators in A such that $\ell(X_{i+1}) \supseteq \ell(X_i)$ for all i. Then $r(\ell(X_i)) \subseteq r(\ell(X_{i+1}))$ for all i. Since A is a ring with the maximum condition on right annihilators, there exists a positive integer n such that $r(\ell(X_i)) = r(\ell(X_n))$ for all $i \geq n$. By **a**, $\ell(X_i) =$

https://doi.org/10.1515/9783110659825-007

$\ell(r(\ell(X_i))) = \ell(r(\ell(X_n))) = \ell(X_n)$ for all $i \geq n$, whence A is a ring with the minimum condition on left annihilators.

If A is a ring with the minimum condition on left annihilators, then it can be similarly proven that A is a ring with the maximum condition on right annihilators.

d. By **c**, A is a ring with the minimum condition on right annihilators. Let $\mathcal{E}$ be the set of all finite subsets of B. The set $\{\ell(E) \mid E \in \mathcal{E}\}$ has a minimal element $\ell(B')$ for some $B' \in \mathcal{E}$. Since $\ell(B' \bigcup b) \subseteq \ell(B')$ for each $b \in B$, we have $\ell(B' \bigcup b) = \ell(B')$, since $\{B' \bigcup b\} \in \mathcal{E}$. Therefore, $\ell(B) = \ell(B')$.

If $0 \neq e = e^2 \in A$, then $r_{eAe}(C) = eAe \cap r_A(C)$ for any subset $C \subseteq eAe$; therefore, eAe is a ring with the maximum condition on right annihilators.

We set $B = \operatorname{Sing}_A A$ and $S = \operatorname{Sing}(A_A)$. Since $r(S^i) \subseteq r(S^{i+1})$ for all i, we have that $r(S^n) = r(S^{n+1})$ for some $n \in \mathbb{N}$. We assume that $S^{n+1} \neq 0$. We can choose $a \in S$ such that $S^n a \neq 0$ and a has a maximal right annihilator among all $x \in S$ such that $S^n x \neq 0$. Let $b \in S$. Since $r(b)$ is an essential right ideal, $r(b) \cap aA \neq 0$. Therefore, there exists an element $t \in A$ such that $at \neq 0$ and $bat = 0$. Then $r(ba) \not\subseteq r(a) \subseteq r(ba)$. By the choice of a, we have $S^n ba = 0$. Then $S^{n+1} a = 0$. Therefore, $a \in r(S^{n+1}) = r(S^n)$. This is a contradiction. Therefore, the ideal S is nilpotent. As is proven above, $\ell(B) = \cap_{i=1}^m \ell(b_i)$ for some $b_1, \ldots, b_m \in B$. Since $\ell(B)$ is the intersection of finitely many essential left ideals, $\ell(B)$ is an essential left ideal. □

7.1.2 Strongly prime and strongly semiprime rings. A ring is said to be *right strongly prime*(see [87]) if each of its nonzero ideals contains a finite subset with the zero right annihilator.

A ring A is said to be *right strongly semiprime*(see [86]) if every ideal X of the ring A that is an essential right ideal, contains a finite subset Y with the zero right annihilator $r(Y)$.

a. Every right strongly prime ring is a right strongly semiprime ring. The direct product of two fields is a strongly semiprime ring that is not strongly prime.

b. If A is a domain or a simple ring, then A is a (right and left) strongly prime ring.

c. If A is a prime ring with the maximum condition on left annihilators, then A is a right strongly prime ring. In particular, every right or left Goldie prime ring is a right strongly prime ring.

d. There exists a strongly prime ring A that is not a right or left finite-dimensional. In particular, A is not a right or left Goldie ring.

e. If A is a right strongly prime ring, then A is a right nonsingular prime ring.

f. There exists a prime ring that is not a right strongly prime ring.

Proof. **a, b.** The assertions are directly verified.

c. The assertion follows from 7.1.1(c).

d. Let A be a free algebra in two variables over a field. Then A is a domain. In particular, A is a right and left strongly prime ring with the maximum condition on right annihilators and with the maximum condition on left annihilators. However, A is not a right or left finite-dimensional ring.

e. It follows from the definition of a right strongly prime ring that A is a prime ring and the ideal Sing A_A contains a finite subset $X = \{x_1, \ldots, x_n\}$ such that $r(X) = r(x_1) \cap \cdots \cap r(x_n) = 0$. All right ideals $r(x_i)$ are essential. Therefore, $r(X)$ is an essential right ideal. Since $r(X) = 0$, this is a contradiction.

f. The assertion follows from **e** and the property that there exist prime rings that are not right nonsingular; e.g., see [115]. □

7.1.3. Let A be a ring, B be its right ideal, AB be the ideal generated by the right ideal B, and let X be a B_A-injective right A-module.

a. X is an $(AB)_A$-injective module.

b. If the ideal AB contains a finite subset $\{y_1, \ldots, y_n\}$ such that $r(\{y_1, \ldots, y_n\}) = 0$, then the module X is injective.

Proof. **a.** Let $\{a_i\}_{i\in I}$ be the set of all elements of the ring A. For any $i \in I$, we denote by f_i a homomorphism from B into Y_i defined by the relation $f_i(b) = a_i b$. Then $AB = \sum_{i\in I} f_i(B)$. By 6.1.1(f), X is an $(AB)_A$-injective module.

b. We have that $r(y_1) \cap \cdots \cap r(y_n) = r(\{y_1, \ldots, y_n\}) = 0$ and $y_i A \cong A_A/r(y_i)$ for any i. Therefore, there exists a monomorphism $A_A \to \oplus_{i=1}^{n} y_i A$. By 6.1.1(e), the module AB is injective. □

7.1.4 Theorem ([179]). Let A be a right strongly prime ring and X a right A-module.

a. If X is injective with respect to some nonzero right ideal of the ring A, then X is an injective module.

b. If X is injective with respect to some right A-module Y that is not an essential extension of a singular module, then X is an injective module.

Proof. **a.** Since the module X is injective with respect to some nonzero right ideal B, it follows from 7.1.3(a) that the module X is injective with respect to the nonzero ideal AB. Since the ring A is right strongly prime, the ideal AB contains a finite subset $\{y_1, \ldots, y_n\}$ such that $r(\{y_1, \ldots, y_n\}) = 0$. By 7.1.3(b), the module X is injective.

b. By 6.2.3(f), there exists a nonzero right ideal B of the ring A such that the module B_A is isomorphic to a submodule of the module Y. Since the module X is injective with respect to the module Y, it follows from 6.1.1(e) that the module X is injective. □

7.1.5 Proposition ([179]). Let A be a right strongly prime ring and let M be a right A-module that is not singular. The following conditions are equivalent:

1) M is an automorphism-extendable module.

2) M is a strongly automorphism-extendable module.

3) $M = X \oplus Y$, where X is an injective singular module, Y is a nonzero nonsingular automorphism-extendable module, and either the module Y is uniform or Y is an injective nonuniform module.

Proof. The implication 3) $\Rightarrow$ 2) is verified with the use of Theorem 6.2.8.

The implication 2) $\Rightarrow$ 1) is true for modules over any ring.

1) $\Rightarrow$ 3). We set $X = \operatorname{Sing} M$. By 7.1.2(e), the right strongly prime ring A is right nonsingular. By 6.2.2(e), the module M/X is nonsingular. Since the ring A is right nonsingular and the module M_A is not singular, M is not an essential extension of the singular of the module X. Therefore, $X \cap Y' = 0$ for some nonsingular submodules Y' in M. By 6.2.4(b), the module X is injective with respect to Y'. By 7.1.4(b), X is an injective module. Therefore, $M = X \oplus Y$, where X is an injective singular module and Y is a nonzero nonsingular automorphism-extendable module. If the module Y is uniform, then the assertion have been proved.

We assume that Y is a nonuniform module. Then there exist two nonzero closed submodules Y_1 and Y_2 in Y such that $Y_1 \cap Y_2 = 0$ and $Y_1 \oplus Y_2$ is an essential submodule in Y. Since Y_1, Y_2 are closed submodules of the nonsingular module Y, the modules M/Y_1 and M/Y_2 are nonsingular. By 6.2.4(a), the module Y_1 is injective with respect to the nonzero nonsingular module Y_2 and the module Y_2 is injective with respect to the nonzero nonsingular module Y_1. By 2), the modules Y_1, Y_2 are injective. Then Y is an essential extension of the injective module $Y_1 \oplus Y_2$. Therefore, $Y = Y_1 \oplus Y_2$, Y is an injective nonuniform module. □

7.1.6. Let A be a right Ore domain, Q be its right classical division ring of fractions, and let M, Y be two nonzero right A-modules.

a. Q_A is the injective hull of the module A_A and for every endomorphism f of the module Q_A, there exists an element $q \in Q$ such that $f(x) = qx$ for all $x \in Q_i$.

b. The module Y is a nonzero uniform nonsingular module if and only if Y is isomorphic to a nonzero submodule of the module Q_A. In the last case, Q_A is the injective hull of the module Y.

c. If the module M is not singular, then the module M is an automorphism-extendable (strongly automorphism-extendable if and only if either M is an injective module which is not singular or $M = X \oplus Y$, where X is an injective singular module and Y is an automorphism-extendable module which is isomorphic to a nonzero submodule in Q_A.

Proof. **a, b.** The assertions are well known.

c. The assertion follows from **b** and 7.1.5. □

In [96, Theorem 6], it is proved that every pseudo-injective nonsingular module over a prime right Goldie ring is injective. In connection to this result, we prove Theorem 7.1.7.

7.1.7 Theorem ([179]). If A is a right strongly prime ring and M is a right A-module, then the following conditions are equivalent.

1) M is an automorphism-invariant module.

2) Either M is a singular automorphism-invariant module or M is an injective module.

Proof. The implication 2) $\Rightarrow$ 1) is obvious.

1) $\Rightarrow$ 2). Automorphism-invariant module M is an automorphism-extendable module. It follows from 7.1.5 that $M = X \oplus Y$, where X is an injective singular module and Y is a nonzero nonsingular automorphism-invariant module and either Y is an injective nonuniform module or the module Y is uniform. It is sufficient to consider the case, where Y is a nonzero automorphism-invariant uniform module. By 6.1.12, Y is a quasi-injective module. By Theorem 7.1.4(b), the module Y is injective. Then the module M is injective. □

7.1.8. Let M be a module, X be a Noetherian submodule in M, and let h be an endomorphism of the module M whose kernel is an essential submodule in M. Then there exists a positive integer $n = n(X, h)$ such that $h^n(X) = 0$.

Proof. We set $X_0 = 0$ and $X_i = X \cap \operatorname{Ker} h^i$, $i = 1, 2, 3 \dots$. Then $X_{i-1} \subseteq X_i$ and $h(X_i) \subseteq h(X_{i-1})$, $i = 1, 2, 3 \dots$. Since X is a Noetherian module, $X_n = X_{n+1}$ for some positive integer n. Let $f \colon X \to M$ be the restriction of the homomorphism h^n to the module X. Since $X_n = X \cap \operatorname{Ker} h_n = \operatorname{Ker} f$, the homomorphism f induces isomorphism $g \colon X/X_n \to g(X/X_n) \subseteq M$. Since $\operatorname{Ker} h$ is an essential submodule of the module M, we have that $\operatorname{Ker} h \cap g(X/X_n)$ is an essential submodule in $g(X/X_n)$. Since $g \colon X/X_n \to g(X/X_n)$ is an isomorphism, $g^{-1}(\operatorname{Ker} h \cap g(X/X_n))$ is an essential submodule in X/X_n. We denote by Y is the complete pre-image in X submodules $g^{-1}(\operatorname{Ker} h \cap g(X/X_n))$ in X/X_n under the action of g. Then

$$\begin{aligned} h^{n+1}(Y) = h(h^n(Y)) = h(f(Y)) &= h(g(g^{-1}(\operatorname{Ker} h \cap g(X/X_n)))) \\ &\subseteq h(\operatorname{Ker} h \cap g(X/X_n)) = 0\,. \end{aligned}$$

Therefore, $Y \subseteq X_{n+1}$ and $Y/X \subseteq X_{n+1}/X_n = 0$. Then $\operatorname{Ker} h \cap g(X/X_n) = g(Y/X) = 0$. Since $\operatorname{Ker} h \cap g(X/X_n)$ is an essential submodule in $g(X/X_n)$, we have $g(X/X_n) = 0$. Therefore, $h^n(X) = f(X) = g(X/X_n) = 0$, which is required. □

7.2 Modules over Hereditary Noetherian Prime Rings

In 7.2.1, we collected some well-known results on right Goldie semiprime rings.

7.2.1 Semiprime Goldie Rings and Their Modules. **a.** A is a right Goldie semiprime ring if and only if A is a right finite-dimensional, right nonsingular semiprime ring, if and only if the set of all essential right ideals of the ring A coincides with the set of all right ideals which contain at least one nonzero divisor, if and only if A is a right finite-dimensional semiprime ring with the maximum condition on left annihilators, if and only if A has the semisimple Artinian right classical ring of fractions.
In this case, A is a ring with the minimum condition on right annihilators and the ring with the minimum condition on left annihilators.

b. Let A be a semiprime right Goldie ring and let M be a right A-module.

We denote by $t(M)$ the set of all elements of M annihilated by some nonzero divisors of the ring A.
A module M is called a *non-torsion* module (resp., *torsion* module; *torsion-free* module) if $t(M) \neq M$ (resp., $t(M) = M$, $t(M) = 0$).

By **a**, the set of all essential right ideals of the ring A coincides with the set of all right ideals which contain at least one nonzero divisor. With the use of this assertion, well known assertions given below are directly verified.

$\operatorname{Sing} M = t(M)$. In particular, M is nonsingular (resp., is singular) if and only if M is a torsion module (resp., a torsion-free module). This property is used without special references.

An essential extension of any singular right A-module is a singular module. Therefore, $\operatorname{Sing} M = G(M)$.

If there exists a submodule X in M such that the modules X and M/X are singular, then the module M is singular.

$\operatorname{Sing} M$ is a closed submodule in M and the module $M/\operatorname{Sing} M$ is nonsingular.

7.2.2 Theorem. Let M be an automorphism-extendable module.

a. If for any endomorphism $h \in \operatorname{End} M$ whose kernel is an essential submodule in M, the endomorphism $1_M - h$ of the module M is an automorphism, then M is a strongly automorphism-extendable module.

b. If for every element $x \in M$ and any endomorphism $h \in \operatorname{End} M$ whose kernel is an essential submodule in M, there exists a positive integer $n = n(x, h)$ such that $h^n(x) = 0$, then M is a strongly automorphism-extendable module.

c. If every cyclic submodule of the module M is a Noetherian module, then M is a strongly automorphism-extendable module.

In particular, every automorphism-extendable right module over right Noetherian ring is a strongly automorphism-extendable module.

Proof. **a.** Let X be a submodule in M and f an automorphism of the module X. We have to prove that f can be extended to an automorphism of the module M. Without loss of generality, we can assume that X is an essential submodule in M. Since M is an automorphism-extendable module, f and f^{-1} can be extended to endomorphisms α and β of the module M, respectively. We denote by h_1 and h_2 endomorphisms $1_M - \beta\alpha$ and $1_M - \alpha\beta$ of the module M, respectively. Since $h_1(X) = 0 = h_2(X)$, we have that $\operatorname{Ker} h_1$, $\operatorname{Ker} h_2$ are essential submodules in M. Since $\beta\alpha = 1_M - h_1$ and $\alpha\beta = 1_M - h_2$, it follows from the assumption that $\beta\alpha$, $\alpha\beta$ are automorphisms of the module M. Therefore, α is an automorphism of the module M.

b. Let $h \in \operatorname{End} M$ and let $\operatorname{Ker} h$ be an essential submodule in M. By **a**, it is sufficient to prove that the endomorphism $1_M - h$ of the module M is an automorphism. We construct a formal series $1_M + \sum_{k=1}^{\infty} h^k$. For every element $x \in M$ there exists a positive integer $n = n(x, h)$ with $h^n(x) = 0$; therefore, $1_M + \sum_{k=1}^{\infty} h^k$ is a correctly defined endomorphism of the module M. It is directly verified that

$$(1_M - h)\left(1_M + \sum_{k=1}^{\infty} h^k\right) = \left(1_M + \sum_{k=1}^{\infty} h^k\right)(1_M - h) = 1_M\,.$$

Therefore, $1_M - h$ is an automorphism of the module M.

c. Let X be an arbitrary cyclic submodule in M and let h be an endomorphism of the module M such that $\operatorname{Ker} h$ is an essential submodule in M. By 7.1.8, there exists a positive integer $n = n(X, h)$ such that $h^n(X) = 0$. By **b**, M is a strongly automorphism-extendable module. □

7.2.3. A ring A is said to be *right bounded* (resp., *left*) if every its essential right (resp., left) ideal contains a nonzero ideal.

a. [119]. Every hereditary Noetherian prime ring is (right and left) bounded or (right and left) primitive and a bounded primitive hereditary Noetherian prime ring is a simple Artinian ring.

b. [58]; also see [63, 25.5.1]. Let A be a hereditary Noetherian prime ring. Then for any nonzero ideal B of the ring A the factor ring A/B is a serial Artinian ring.

c. If a hereditary Noetherian domain A is left invariant, then the domain A is right invariant and for any nonzero ideal B of the ring A, the factor ring A/B is the finite direct product of invariant uniserial Artinian rings.

Proof. **a.** See [119].

b. See [58]; also see [63, 25.5.1].

c. Without loss of generality, we can assume that A is not a division ring. Since every left invariant, left primitive ring is a division ring, A is not a left primitive ring. By **a**, A

is a right bounded ring. We have to prove that an arbitrary nonzero proper right ideal C of the ring A is an ideal. Since A is a right Ore domain, C is an essential right ideal. Since A is a right bounded ring, C contains a nonzero proper ideal B of the ring A. By **b**, A/B is a serial Artinian ring. In addition, A/B is a left invariant ring. Therefore, A/B is a direct product of finitely many left invariant, right and left uniserial, right and left Artinian rings. It is directly verified that every left invariant, right and left uniserial, right and left Artinian ring is a right invariant ring. Therefore, A/B is a right invariant ring, whence C/B is an ideal in A/B. In addition, B is an ideal in A. Therefore, C is an ideal in A. □

7.2.4 Theorem. If A is a hereditary Noetherian prime ring and M is an A-module with nonzero annihilator $r(M)$, then the following conditions are equivalent.

1) M is an automorphism-extendable module.

2) M is a strongly automorphism-extendable module.

3) M is an automorphism-invariant module.

4) M is a quasi-injective module.

Proof. The implications 4) ⇒ 3) ⇒ 2) ⇒ 1) are true for modules over any ring.

1) ⇒ 4). By 7.2.3(b), the factor ring $A/r(M)$ is a serial Artinian ring. Since M is an automorphism-extendable A-module, M is an automorphism-extendable $A/r(M)$-module. By Theorem 6.1.15, M is a quasi-injective $A/r(M)$-module. Therefore, M is a quasi-injective A-module. □

7.2.5. Let M be a module and $M = \sum_{i\in I} M_i$, where all M_i are essential quasi-injective submodules in M. Then M is a quasi-injective module.

Proof. Let Q be the injective hull of the module M. Since all M_i are essential submodules in M, all M_i are essential submodules of the injective of the module Q. Therefore, Q is the injective hull of the module M_i for any $i \in I$. Let f be an endomorphism of the module Q. Since all modules M_i are quasi-injective, $f(M_i) \subseteq M_i$ for any $i \in I$. Then

$$f(M) = f\left(\sum_{i\in I} M_i\right) = \sum_{i\in I} f(M_i) \subseteq \sum_{i\in I} M_i = M\,.$$

Therefore, the module M is quasi-injective. □

7.2.6. Let A be a hereditary Noetherian prime ring and M an automorphism-extendable A-module.

a. If B is a nonzero ideal of the ring A and $X = \{m \in M \mid mB = 0\}$, then X is a quasi-injective module.

b. Let $\{B_i\}_{i\in I}$ be some set of nonzero ideals of the ring A and $X_i = \{m \in M \mid mB_i = 0\}$, $i \in I$. If $M = \sum_{i\in I} X_i$ and all X_i are essential submodules in M, then M is a quasi-injective module.

Proof. **a.** Let f be an endomorphism of the module M. Then

$$f(X)B = f(XB) = f(0) = 0\,, \quad f(X) \subseteq X\,.$$

Therefore, X is a fully invariant submodule of the automorphism-extendable module M. Therefore, X is an automorphism-extendable A-module with nonzero annihilator. By Theorem 2.27, X is a quasi-injective module.

b. By **a**, all modules M_i are quasi-injective. By 7.2.5, the module M is quasi-injective. □

In [96, Theorem 5], it is proved that every pseudo-injective torsion module over a bounded hereditary Noetherian prime ring is quasi-injective. In connection to this result, we will prove Theorem 7.2.7.

7.2.7 Theorem ([176]). If A is a bounded hereditary Noetherian prime ring and M is a right A-module, then the following conditions are equivalent.

1) M is an automorphism-extendable module.

2) M is a strongly automorphism-extendable module.

3) Either M is a quasi-injective singular module or M is an injective module which is not singular, or $M = X \oplus Y$, where X is an injective singular module and Y is a nonzero automorphism-extendable uniform nonsingular module.

Proof. The implication 3) $\Rightarrow$ 2) follows from 7.1.5 and the property that every quasi-injective module is a strongly automorphism-extendable.

The implication 2) $\Rightarrow$ 1) is true for modules over any ring.

1) $\Rightarrow$ 3). If the module M is not singular, it follows from 7.1.5 that $M = X \oplus Y$, where X is an injective singular module, Y is a nonzero nonsingular automorphism-extendable module, and either the module Y is uniform or Y is an injective nonuniform module.

Now we assume that M is a singular automorphism-extendable module. Let $\{B_j\}_{j\in J}$ be the set of all proper of invertible ideals of the ring A, $\{P_i\}_{i\in I}$ be the set of all maximal elements of the set $\{B_j\}_{j\in J}$. For any $i \in I$, we denote by M_i submodule in M formed by all elements of M which are annihilated by some degree of the ideal P_i.

In [146], it is proved that every singular module M has the following two properties.

i) For any submodule X of M, it is true that $X = \oplus_{i\in I} X_i$, where $X_i = X \cap M_i$, $i \in I$ and $\mathrm{Hom}(X_i, X_j) = 0$ for any distinct subscripts $i, j \in I$.

ii) $Mi, k \subseteq Mi, k+1$ for every $k \in \mathbb{N}$, $M_i = \cup_{k\in\mathbb{N}} Mi, k = \sum_{k\in\mathbb{N}} Mi, k$, and Mi, k is an essential fully invariant submodule in M_i for every $k \in \mathbb{N}$.

With the use of the assertion i), the following two properties are directly verified.

iii) The module M is automorphism-extendable if and only if all modules M_i are automorphism-extendable.

iv) The module M is quasi-injective if and only if all modules M_i are quasi-injective.

It follows from iii) that all M_i are automorphism-extendable modules. By iv), it is sufficient to prove that all M_i are quasi-injective modules.

Next, we fix $i \in I$ and denote by Mi, k, $k \in \mathbb{N}$, the submodule in M_i annihilated by the ideal P_i^k. By ii), $Mi, k \subseteq Mi, k+1$ for every $k \in \mathbb{N}$, $M_i = \cup_{k\in\mathbb{N}} Mi, k = \sum_{k\in\mathbb{N}} Mi, k$ and Mi, k is an essential fully invariant submodule in M_i for every $k \in \mathbb{N}$. The fully invariant submodules Mi, k of the automorphism-extendable module M_i are automorphism-extendable modules with nonzero annihilators. By Theorem 7.2.4, every essential submodule Mi, k in M_i is a quasi-injective module. By 7.2.5, M is a quasi-injective module. □

7.2.8 Theorem. If A is a bounded hereditary Noetherian prime ring and M is a right A-module, then the following conditions are equivalent.

1) M is an automorphism-invariant module.

2) M is a quasi-injective module.

3) Either M is a quasi-injective singular module or M is an injective module which is not singular.

Proof. The implications 3) ⇒ 2) ⇒ 1) are true for modules over any ring.

1) ⇒ 3). By Theorem 7.2.7, either M is a quasi-injective singular module or M is an injective module which is not singular, or $M = X \oplus Y$, where X is an injective singular module and Y is a nonzero automorphism-invariant uniform nonsingular module.

It is sufficient to consider only modules $M = X \oplus Y$, where X is an injective singular module and Y is a nonzero automorphism-invariant uniform nonsingular module. By 6.1.12, Y is a quasi-injective module. By Theorem 7.1.4(b), Y is an injective module. Since $M = X \oplus Y$, we have that M is an injective module. □

7.2.9 Remark. Let A be a bounded hereditary prime ring. Quasi-injective A-modules are described in [145]. Therefore, Theorem 7.2.8 completely describes all automorphism-invariant A-modules and Theorem 7.2.7 describes all automorphism-extendable A-modules up to the description of automorphism-extendable uniform nonsingular modules.

7.2.10. Let A be an invariant hereditary domain and Q the division ring of fractions of the domain A.

a. If M is an arbitrary submodule of any cyclic singular A-module, then $M = M_1 \oplus \cdots \oplus M_n$, where all M_i are uniserial modules of finite length. In addition, $f(m) \in mA$ for any element $m \in M$ and each homomorphism $f: mA \to M$.

b. If X is any nonzero submodule in Q_A and M is a submodule in Q_A such that $X \subseteq M$ and M/X is a finitely generated module, then $\bar{f}(\bar{m}) \in \bar{m}A$ for any element $\bar{m} \in M/X$ and each homomorphism $\bar{f}: \bar{m}A \to M/X$.

c. If X is any nonzero submodule in Q_A and Y is a submodule in Q_A such that $X \subseteq Y$, then $\bar{f}(\bar{y}) \in \bar{y}A$ for any element $\bar{y} \in Y/X$ and each homomorphism $\bar{f}: \bar{y}A \to Y/X$.

d. Every submodule Y of the module Q_A is a strongly automorphism-extendable module and a strongly endomorphism-extendable module.

e. Every uniform nonsingular A-module is a strongly automorphism-extendable module and a strongly endomorphism-extendable module.

Proof. **a.** The assertion follows from 7.2.3(c).

b. Since M/X is a finitely generated module, there exists a finitely generated submodule N in M such that $M = N + X$. There exists a natural isomorphism $g: M/X \to N/(N \cap X)$. Since N is a finitely generated A-submodule of the division ring of fractions Q of the domain A, it follows from the left-right symmetrical analogue of 1.3.2(a) that there exists a monomorphism $h: N \to A_A$. Then the module $h(N)/h(N \cap X)$ is isomorphic to a submodule of the cyclic singular module $A_A/h(N \cap X)$. By **a**, $f(m) \in mA$ for any element $m \in h(N)/h(N \cap X)$ and each homomorphism $f: mA \to h(N)/h(N \cap X)$. Since there exists a natural isomorphism $M/X \to N/(N \cap X)$, we have that $\bar{f}(\bar{m}) \in \bar{m}A$ for any element $\bar{m} \in M/X$ and each homomorphism $\bar{f}: \bar{m}A \to M/X$.

c. Let $\bar{y} = y + X \in Y/X$, where $y \in Y$. We set $M = X + yA$. Then M/X is a cyclic module. By **b**, $\bar{f}(\bar{y}) \in \bar{y}A$ for every homomorphism $\bar{f}: \bar{y}A \to Y/X$.

d. We prove that Y is a strongly endomorphism-extendable module. Let M be a nonzero submodule in Y, X be an essential submodule in M, and let $g: M: Y$ be a homomorphism with $g(X) \subseteq X$. Since the module Q_A is injective and $g(X) \subseteq X$, the homomorphism g can be extended to an endomorphism f of the module Q_A and $f(X) \subseteq X$. Then f induces the endomorphism $\bar{f}$ of the module Q/X. By **d**, $\bar{f}(Y/X) \subseteq Y/X$. Therefore, Y is a strongly endomorphism-extendable module.

It can be similarly proved that Y is a strongly automorphism-extendable module.

e. The assertion follows from **e** and 7.1.6(b). □

7.2.11 Theorem. Let A be an invariant hereditary domain, Q be the division ring of fractions of the domain A, and let M be a right A-module. The following conditions are equivalent.

1) M is an automorphism-extendable module.

2) M is a strongly automorphism-extendable module.

3) M is an endomorphism-extendable module.

4) M is a strongly endomorphism-extendable module.

5) Either M is a quasi-injective singular module or M is an injective module which is not singular, or $M = X \oplus Y$, where X is an injective singular module and the module Y is isomorphic to nonzero submodule in Q_A.

Proof. The implications 5) ⇒ 4) and 5) ⇒ 2) follow from 7.2.10(d).

The implications 4) ⇒ 3)⇒ 1) and 2) ⇒ 1) are true for modules over any ring.

The implication 1) ⇒ 5) follows from Theorem 7.2.7 and 7.1.6(b). □

7.2.12 Proposition. Let A be a principal right ideal domain and U its group of invertible elements. The following conditions are equivalent.

1) A_A is an automorphism-extendable module.

2) A_A is a strongly automorphism-extendable module.

3) $aU \subseteq Ua$ for any element $a \in A$.

Proof. The implication 2) ⇒ 1) is directly verified.

1) ⇒ 3). We have to prove that $au \in Ua$ for any elements $a \in A$ and $u \in U$. Without loss of generality, we can assume that $a \neq 0$. We denote by φ a mapping from aA into A such that $\varphi(ab) = aub$ for any element $b \in A$. Since $aA = auA$ and A is a domain, φ is an automorphism of the module aA. Since A_A is an automorphism-extendable module, the automorphism φ can be extended to some endomorphism f of the module A_A. We set $v = f(1) \in A$. Since $vA = A$ and A is a domain, $v \in U$. Then $va = f(a) = au$ and $aU \subseteq Ua$.

3) ⇒ 2). Let X be a submodule in A_A and φ its automorphism. Since A is a principal right ideal domain, $X = aA$ for some element $a \in X$. Without loss of generality, we can assume that $a \neq 0$. Since $aA = \varphi(a)A$, there exist elements $u, w \in A$ such that $\varphi(a) = au$ and $a = auw$. Since A is a domain, $1 = uw$ and $u \in U$. By assumption, $aU \subseteq Ua$. Therefore, $au = va$ for some $v \in U$. We denote by f the automorphism of the module A_A such that $f(b) = vb$ for all $b \in A$. Then $f(a) = va = au = \varphi(a)$. Therefore, the automorphism f is an extension of the automorphism φ. □

7.2.13 Proposition. Let D be a noncommutative division ring. Then $D[x]$ is a principal right (left) ideal domain, which is not an automorphism-extendable right or left $D[x]$-module. In addition, if the division ring D is finite-dimensional over its center F, then $D[x]$ is a bounded hereditary Noetherian prime ring.

Proof. It is well known that $D[x]$ is a principal right (left) ideal domain and its group of invertible elements U coincides with the multiplicative group of the division ring D. In particular, $D[x]$ is a hereditary Noetherian prime ring. We assume that $D[x]_{D[x]}$ is an automorphism-extendable module. By assumption, $dd_1 \neq d_1d$ for some nonzero elements d, d_1 of the division ring D. By Proposition 7.2.12, $(d+x)d_1 \subseteq Ud$. In addition, $U = D \setminus 0$. Therefore, $(d + x)d_1 = d_2(d + x)$ for some element d_2 of the division ring D. Then $d_1x = d_2x$ and $dd_1 = d_2d$. Therefore, $dd_1 = d_1d$. This is a contradiction. We similarly obtain that the module ${}_{D[x]}D[x]$ is not automorphism-extendable.

We assume that D has finite dimension over its center F. It is well known that for any polynomial $f \in D[x]$, there exists a polynomial $g \in D[x]$ such that fg is a nonzero polynomial from $F[x]$; e.g., see [112, 16.9]. Then fg is a nonzero central element of the domain $D[x]$, contained in the principal right (left) ideal domain $fD[x]$. Therefore, $D[x]$ is a bounded hereditary Noetherian prime ring. □

7.2.14 Example. Let $\mathbb{H}$ be the division ring of Hamiltonian quaternions and let $\mathbb{R}$ be the field of real numbers. Since the noncommutative division ring $\mathbb{H}$ is finite-dimensional over its center $\mathbb{R}$, it follows from Proposition 7.2.13 that $\mathbb{H}[x]$ is a bounded principal right (left) ideal domain that is not an automorphism-extendable right or left $D[x]$-module. In particular, $\mathbb{H}[x]$ is a bounded hereditary Noetherian prime ring that is not automorphism-extendable right or left $D[x]$-module.

7.2.15 The completion of the proof of Theorems 7A and 7B. Theorems 7A and 7B follows from Theorems 7.1.7 and 7.2.11.

7.2.16 Remark. Since the ring $\mathbb{Z}$ is an invariant hereditary domain, Theorem 7B also describes Abelian groups, which are (strongly) automorphism-extendable or endomorphism-extendable; see [130].

7.2.17 Theorem ([111, Theorem 3.4, Theorem 3.8]). Let A be a ring and $G = G(A_A)$. All nonsingular quasi-injective right A-modules are injective if and only if the ring A/G is right strongly semiprime.

7.2.18 Open question. Is it true that every endomorphism-extendable module is strongly endomorphism-extendable?

8 Endomorphism-extendable modules and rings

The main results of this section are Theorems 8A and 8B.

8A Theorem (Tuganbaev [167]). A ring A is a right endomorphism-extendable, right nonsingular ring if and only if $A = B \times C$, where B is a right injective regular ring, C is a left invariant, reduced Baer ring and C is a right completely integrally closed subring of its maximal right rings of fractions Q.

8B Theorem (Tuganbaev [162]). A ring A is right (left) Noetherian ring such that all cyclic right (left) modules are endomorphism-extendable if and only if $A = A_1 \times \cdots \times A_n$, where A_i is either a simple Artinian ring or a uniserial Artinian ring, or an invariant hereditary Noetherian domain, $i = 1, \ldots, n$.

1) M is an automorphism-extendable (strongly automorphism-extendable) module.

2) M is an endomorphism-extendable (strongly endomorphism-extendable) module.

3) Either M is a quasi-injective singular module or M is an injective module that is not singular, or $M = X \oplus Y$, where X is an injective singular module and the module Y is isomorphic to nonzero submodule in Q_A.

Remark. For the completion of the proof of Theorems 8A and 8B, see 8.3.13.

8.1 Strongly endomorphism-extendable modules

8.1.1 Quasicontinuous, *CS* and C_3 modules. A module M is called a *CS module* or a C_1 *module* if every submodule in M is an essential submodule of some direct summand of the module M.

A module M is called a C_3 *module* if $X \oplus Y$ is a direct summand of M for any direct summands X and Y of M such that $X \cap Y = 0$.

A module M is said to be *quasicontinuous* [100] or *π-injective* if the following equivalent conditions hold.

1) Every idempotent endomorphism of any submodule in M can be extended to an endomorphism of the module M.

2) Every idempotent endomorphism of any submodule in M can be extended to an idempotent endomorphism of the module M.

3) M is an *idempotent-invariant* module, i.e., $\alpha(M) \subseteq M$ for every idempotent endomorphism α of the injective hull of the module M.

4) M is a CS module and a C_3 module.

https://doi.org/10.1515/9783110659825-008

5) $M = \oplus_{i \in I}(M \cap Q_i)$ for any direct decomposition $Q = \oplus_{i \in I} Q_i$ of the injective hull Q of the module M.

6) For any submodule $X = X_1 \oplus X_2 \oplus \cdots \oplus X_n$ of the module M, there exists a direct decomposition $M = M_1 \oplus \oplus M_2 \cdots \oplus M_n \oplus Y$ of the module M such that M_i is an essential extension of the module X_i, $i = 1, 2, \ldots n$.

Proof. The equivalences 2) ⇔ 3) ⇔ 4) are proven in [100] and are well known; also see [166].

The equivalences 3) ⇔ 5) and 4) ⇔ 6) are directly verified; also see [166].

The implication 2) ⇒ 1) is obvious.

The implication 1) ⇒ 3) follows from 6.1.4(c). □

8.1.2 Quasi-injective, endomorphism-extendable and quasicontinuous modules. a. If a module M is quasi-injective, then it is clear that M is an endomorphism-extendable module. In addition, $\mathbb{Z}_{\mathbb{Z}}$ is an endomorphism-extendable module, which is not quasi-injective.
Indeed, it is directly verified that $\mathbb{Z}_{\mathbb{Z}}$ is an endomorphism-extendable module. The module $\mathbb{Z}_{\mathbb{Z}}$ is not quasi-injective, since the $\mathbb{Z}$-module homomorphism $2\,\mathbb{Z}_{\mathbb{Z}} \to \mathbb{Z}_{\mathbb{Z}}$, $2z \to z$ cannot be extended to an endomorphism of the module $\mathbb{Z}_{\mathbb{Z}}$.

b. If M is an endomorphism-extendable module, then M is a quasicontinuous module by 8.1.1. In addition, by 3.3.7(m), there exist quasicontinuous modules that are not endomorphism-extendable.

c. Every fully invariant submodule of any quasi-injective (resp., endomorphism-extendable; automorphism-extendable; quasicontinuous) module is a quasi-injective (resp., endomorphism-extendable; automorphism-extendable; quasicontinuous) module.

d. A right A-module M is quasicontinuous if and only if M is a quasicontinuous $A/r(M)$-module, where $r(M)$ is the annihilator of the module M.

e. Right quasicontinuous domains coincide with right Ore domains (right uniform domains).

f. All submodules of a uniform module are quasicontinuous finite-dimensional modules.

g. Finite-dimensional modules coincide with modules of finite Goldie dimension and Goldie dimension are uniquely defined.

h. Quasicontinuous finite-dimensional modules are finite direct sums of uniform modules.

i. Right finite-dimensional domains coincide with right Ore domains.

The assertions of 8.1.2 are well known; most of them are directly verified.

In [116, Theorem 12], it is proven that every automorphism-invariant module M satisfies property C_3. In connection to this result, we prove Theorem 8.1.3, where the proof of the first assertion is similar to the proof of [116, Theorem 12].

8.1.3 Theorem ([176]). Let M be a strongly automorphism-extendable module. Then M is a C_3 module. In addition, if M is a CS module, then M is a quasicontinuous module.

Proof. Let $M = A \oplus A' = B \oplus B'$ and $A \cap B = 0$. We have to prove that $A \oplus B$ is a direct summand of M. Let $\pi\colon M \to A'$ be the projection with kernel A. There exists a submodule C in M such that $(A+B)\cap C = 0$ and $A\oplus B\oplus C$ is an essential submodule in M. We set $D = B\oplus C$. Then $A\oplus D = A\oplus \pi D$ and $\pi|_D\colon D \to \pi D$ is an isomorphism. Therefore, $1_A\oplus\pi_D\colon A\oplus D \to A\oplus\pi D$ is an automorphism of the module $A\oplus D$. Since M is a strongly automorphism-extendable module, $1_A \oplus \pi_D$ can be extended to an automorphism f of the module M. Since B is a direct summand of the module M, then $\pi B = fB$ is a direct summand of the module M. Therefore, πB is a direct summand of the module A', whence $A \oplus B = A \oplus \pi B$ is a direct summand of the module M. Therefore, M is a C_3 module.

Now the second assertion follows from the definition of a quasicontinuous module. □

8.1.4. In connection to Theorem 8.1.3, we remark that $\mathbb{Z}_{\mathbb{Z}}$ is a strongly automorphism-extendable quasicontinuous module which is not automorphism-invariant.

8.1.5. Let A be a ring, Q be a right A-module, and let M be an essential submodule in Q.

a. Let X be a module, $f\colon X \to Q$ be a homomorphism, and let us have a homomorphism $g\colon X \to M$ such that f coincides with g on $f^{-1}(M)$. Then $f(X) \subseteq M$.

b. If f is an endomorphism of the module Q and there exists an endomorphism g of the module M such that f coincides with g on $M \cap f^{-1}(M)$, then $f(M) \subseteq M$.

c. For any module homomorphism $f\colon X \to Q$, it is true that $f^{-1}(M)$ is an essential submodule in X.

d. Q/M is a singular module.

e. If M is a nonsingular module, then Q is a nonsingular module and the kernel of any nonzero module homomorphism $f\colon X \to Q$ is not an essential submodule in X. In particular, Q does not have nonzero endomorphisms with essential kernels.

Proof. **a.** We assume that $m = (f - g)(x) \in M \cap (f - g)(M')$, where $x \in X$. Then

$$f(x) = (f-g)(x) + g(x) = m + g(x) \in M\,, \quad x \in f^{-1}(M)\,,$$
$$m \in (f-g)(f^{-1}(X)) = 0\,, \quad M \cap (f-g)(X) = 0\,.$$

Since Q is an essential extension of the module M, we have $(f - g)(X) = 0$. Therefore, $f(X) = g(X) \subseteq M$.

b. The assertion follows from **a** for $X = M$.

c. Let Y be a nonzero submodule in X. If $f(Y) = 0$, then $0 \neq Y \subseteq f^{-1}(0) \subseteq Y \cap f^{-1}(M)$. We assume that $f(Y) \neq 0$. Then $M \cap f(Y) \neq 0$, whence $0 \neq f^{-1}(M \cap f(Y)) \subseteq f^{-1}(M)$ and $Y \cap f^{-1}(M) \neq 0$.

d. Let $q \in Q$ and let $f\colon A_A \to Q$ be a homomorphism such that $f(a) = qa$ for all $a \in A$. By **c**, $f^{-1}(M)$ is an essential right ideal of the ring A. Since $f^{-1}(M)$ is the annihilator of the element $q + X \in Q/X$, we have that Q/M is a singular module.

e. Since $M \cap \operatorname{Sing} Q = 0$ and Q is an essential extension of the module M, the module Q is nonsingular. If $f \in \operatorname{Hom}(X, Q)$ and X is an essential extension of the module $\operatorname{Ker} f$, it follows from **d** that $\operatorname{Sing}(N/\operatorname{Ker} f) = N/\operatorname{Ker} f \cong f(N)$, whence $f(N) = \operatorname{Sing}(f(N)) \subseteq \operatorname{Sing} M = 0$, f is a nonzero homomorphism. □

8.1.6 Strongly endomorphism-extendable modules. We recall that the module M is said to be *strongly endomorphism-extendable* if for any submodule X in M, every homomorphism $X \to M$, which maps into itself some essential submodule of X, can be extended to a homomorphism $M \to M$.

If M is a module with of the injective hull Q, then the following conditions are equivalent.

1) M is a strongly endomorphism-extendable module.

2) $f(M) \subseteq M$ for any endomorphism f of the module Q that maps into itself some essential submodule of the module M.

3) M is an endomorphism-extendable module and $h(M) \subseteq M$ for any endomorphism h of the module Q such that $\operatorname{Ker} h$ is an essential submodule in Q.

Proof. 1) ⇒ 2). Let X be an essential submodule of the module M and let f be an endomorphism of the module Q such that $f(X) \subseteq X$. We set $Y = M \cap f^{-1}(M)$. Then $X \subseteq Y$, f induces the homomorphism $g_1\colon Y \to M$, and $g_1(X) = f(X) \subseteq X$. Since M is a strongly endomorphism-extendable module, the homomorphism g_1 can be extended to some endomorphism g of the module M. By 8.1.5(b), $f(M) \subseteq M$.

2) ⇒ 3). We denote by X the essential submodule $M \cap \operatorname{Ker} h$ of the module M. Since $h(X) = 0 \subseteq X$, it follows from) that $h(M) \subseteq M$.

Let X be an essential submodule in M and let g_1 be an endomorphism of the module X. Since the module Q is injective, g_1 can be extended to an endomorphism f of the module Q. By 2), f induces the endomorphism g of the module M, which is an extension of the endomorphism g_1. By 6.1.6(c), M is an endomorphism-extendable module. □

8.1.7. Every nonsingular endomorphism-extendable module M is strongly endomorphism-extendable.

Proof. By 8.1.5(e,), the injective hull Q of the nonsingular module M does not have nonzero endomorphisms with essential kernels. Therefore, the assertion follows from 8.1.6. □

8.1.8. A module is said to be *locally Noetherian* if each of its cyclic submodules is Noetherian.

An endomorphism f of the module X is said to be *locally nilpotent* if $X = \bigcup_{n=1}^{\infty} \operatorname{Ker} f^n$, i.e., if for any $x \in X$, there exists a positive integer n with $f^n(x) = 0$. In this case, for any finitely generated submodules Y in X, there exists a positive integer n with $f^n(Y) = 0$.

Let A be a ring, M be an endomorphism-extendable right A-module, and let Q be the injective hull of the module M.

a. If every essential submodule in M is fully invariant in M, then M is a strongly endomorphism-extendable module.

b. If for any essential submodule N in M, the factor module M/N is semi-Artinian, then M is a strongly endomorphism-extendable module.

c. If for any $m \in M$ and each endomorphism $g \in \operatorname{End} M$ with essential in M kernel, there exists a positive integer n such that $g^{n+1}(m) \in \sum_{i=0}^{n} g^i(m)A$, then the module M is strongly endomorphism-extendable.

d. If every endomorphism $g \in \operatorname{End} M$ with an essential M kernel is locally nilpotent, then the module M is strongly endomorphism-extendable.

e. If M is a local Noetherian module, then M is a strongly endomorphism-extendable module.

Proof. Let h be an endomorphism of the injective hull Q of the module M such that $\operatorname{Ker} h$ is an essential submodule in Q. By 8.1.6, it is sufficient to prove that $h(M) \subseteq M$ in all considered cases **a–e.**

Let $P \equiv M \cap h^{-1}(M)$ and $N \equiv \{m \in M \mid h^n(M) \subseteq M \text{ for all } n \in \mathbb{N}\}$. Since $N \supseteq M \cap \operatorname{Ker} h$, we have that Q and M are essential extensions of the module N. In addition, N is the largest submodule in M with property $h(N) \subseteq N$. Since M is an endomorphism-extendable module and $h(N) \subseteq N$, we have that $(h - g)(N) = 0$ for some $g \in \operatorname{End} M$. Since $g(M \cap \operatorname{Ker} h) = 0$, we have $g \in \operatorname{sg} M$. We set $\overline{M} = M/N$. If $\overline{M} = 0$, then $M = N$, $h(M) \subseteq M$ and the assertion have been proven in this case.

Now we assume that $\overline{M} \neq 0$. Then $(h - g)(N) = 0$ and a homomorphism $t \colon \overline{M} \to Q$ is correctly defined by the rule $t(x + N) = (h - g)(x)$. Let $\overline{V} \equiv t^{-1}(N) \subseteq \overline{M}$ and let V be a submodule in M such that $V \supseteq N$ and $V/N = \overline{V}$. Since Q is an essential extension of the module M, it follows from 6.2.1(b) that $\overline{M}$ is an essential extension of the module $\overline{V}$. Therefore, $\overline{V} \neq 0$. In addition, $h(V) \subseteq t(V) + g(V) \subseteq N + g(V) \subseteq M$. Since $V \supseteq N$, we have that M is an essential extension of V.

a. By assumption, $g(V) \subseteq V$. Therefore, $h(V) \subseteq N + g(V) \subseteq N + V = V$. Since N is the largest submodule in M with property $h(N) \subseteq N$, we have $V = N$. Therefore, $\overline{V} = 0$. This is a contradiction.

b. Since $h(V) \subseteq M$ and $h(N) \subseteq N$, we have that h induces the homomorphism $\overline{h}\colon \overline{V} \to \overline{M}$. Since M is an essential extension of the module V, it follows from the assumption that the nonzero module $\overline{M}$ is an essential extension of its nonzero socle $\overline{S} = S/N$, where $N \in \text{Lat}(S)$. Since $\overline{V}$ is an essential submodule in $\overline{M}$, we have $\overline{S} \subseteq \overline{V}$, whence $\overline{h}(\overline{S}) \subseteq \overline{S}$. Therefore, $h(S) \subseteq S$. Since N is the largest submodule in M with property $h(N) \subseteq N$, we have $S = N$. Therefore, $\overline{S} = 0$. This is a contradiction.

c. Let $m \in M \setminus N$, $0 \neq \overline{m} = m + N \in \overline{M}$. By assumption, $g^{n+1}(m) \in \sum_{i=0}^{n} g^i(m)A$ for some n. Since $(h^i - g^i)(N) = 0$ for all i, homomorphisms $t_i\colon \overline{m}A \to E$ are correctly defined by the rule $t_i(\overline{m}x) = (h^i - g^i)(mx)$, $x \in A$. Let $\overline{V}_i \equiv t_i^{-1}(N) \subseteq \overline{m}A$ and $\overline{W} \equiv \overline{V}_1 \cap \ldots, \cap \overline{V}_n \subseteq mA$. Since Q is an essential extension of the module M, we have that $\overline{m}A$ is an essential extension of each of the modules $\overline{V}_i$. Therefore, $\overline{m}A$ is an essential extension of the module $\overline{W}$. Then $0 \neq \overline{m}a \in \overline{W}$ for some $a \in A$. Therefore, $h^i(ma) = t_i(ma) + g^i(ma) \in M$ for all $i = 1, \ldots, n$ and $h^{n+1}(ma) \in \sum_{i=0}^{n} h^i(ma)A \in M$. Then $h^{n+j}(ma) \in M$ for all $j \geq 1$. Therefore, $h^k(ma) \in M$ for all $k \geq 1$, $ma \in N$, $\overline{m}a = 0$. This is a contradiction.

d. The assertion follows from **c.**

e. Let X be an arbitrary cyclic submodule in M. By 7.1.8, there exists a positive integer $n = n(X, h)$ such that $h^n(X) = 0$. Therefore, the assertion follows from **d.** □

8.1.9 Proposition. Let M be a strongly endomorphism-extendable module and let S be the sum of the images of all endomorphisms of the module M with essential kernels.

a. The ideal sg M is contained in the Jacobson radical $J(\text{End}\, M)$.

b. If M is an essential extension of some quasi-injective module S, then M is a quasi-injective module. In particular, if M is an essential extension of a semisimple module, then M is a quasi-injective module.

c. If M is an essential extension of the module S, then M is a quasi-injective module and $M = G(M)$.

d. If M is an indecomposable module, then M is a uniform module and either M is a quasi-injective module and $M = G(M)$ or every nonzero endomorphism of the module M is a monomorphism and End M is a domain.

Proof. Let Q be the injective hull of the module M.

a. Let $h \in \text{sg}\, M$. Then $h(X) = 0$ for some essential submodule X in M. Since sg M is an ideal of the ring End M, it is sufficient to prove that $1_M - h$ is an automorphism of the module M. Since the module Q is injective, we have that $1_M - h$ can be extended to an endomorphism f of the module Q which acts identically on the essential submodule X of the injective module Q. Therefore, $f(Q)$ is an injective essential submodule of the module Q. Then f is an automorphism of the injective module Q. The inverse automorphism f^{-1} also acts identically on the essential submodule X of the module Q. Therefore, $1_Q - f^{-1} \in \operatorname{sg} Q$. By 8.1.6, $(1_Q - f^{-1})(M) \subseteq M$. Then $f^{-1}(M) \subseteq M$. Therefore, $M \subset f(M) = (1_M - h)(M) \subseteq M$. Then $1_M - h$ is an automorphism of the module M.

b. It is sufficient to prove that $f(M) \subseteq M$ for any endomorphism f of the module Q. Since M is an essential extension of the module X, we have that Q is the injective hull of the quasi-injective module X. Since X is a quasi-injective module, $f(X) \subseteq X$. By 8.1.6, $f(M) \subseteq M$. The second assertion follows from the first assertion and the property that every semisimple module is quasi-injective.

c. By 8.1.6, $h(M) \subseteq M$ for any $h \in \operatorname{sg} Q$. Therefore, $\sum_{h \in \operatorname{sg} Q} h(Q) = S$ is an essential submodule in M. Then Q is the injective hull of the module S. By 6.2.1(d), $S \subseteq \operatorname{Sing} Q$ and S is a fully invariant invariant submodule in Q. Therefore, $M = G(M)$ and S is a quasi-injective module. By **b**, M is a quasi-injective module.

d. Since M is an indecomposable quasicontinuous module, M is a uniform module.

If $S \neq 0$, then M is an essential extension of the module S and by **b**, M is a quasi-injective module and $M = G(M)$.

We assume that $S = 0$, i.e., the kernel of each nonzero endomorphism of the module M is not essential in M. Since M is a uniform module, every nonzero endomorphism of the module M is a monomorphism. Therefore, $\operatorname{End} M$ is a domain. □

8.2 Endomorphism-extendable rings

8.2.1 Baer rings. A ring A is called a *Baer* ring if the following equivalent conditions hold.

1) For any subset X in A, there exists an idempotent $c \subset A$ with $r(X) = eA$.

2) For any subset Y in A, there exists an idempotent $f \in A$ with $\ell(Y) = Af$.

a. Every Baer ring is (right and left) nonsingular.

b. Every right nonsingular, right quasicontinuous ring A is a Baer ring. In particular, A is left nonsingular.

c. If M is a quasicontinuous module, then there exists a natural ring isomorphism from the ring $\operatorname{End} M/\operatorname{sg} M$ onto the direct product of a right injective regular ring and a reduced ring.

d. Any right nonsingular, right quasicontinuous ring is the direct product of a right injective regular ring and a right quasicontinuous reduced Baer ring.

Proof. The equivalence 1) ⇔ 2) from the definition of a Baer ring is well known and it is directly verified. For example, let's prove the implication 1) ⇒ 2). By 1), there exists an idempotent $e \in A$ such that $r(\ell(Y)) = eA$. Then

$$\ell(Y) = \ell(r(\ell(Y))) = \ell(eA) = A(1 - e)\,.$$

The assertions **a**is **a**d are well known and the assertion **a** is directly verified.

The assertion **b** is known; e.g., see [57, 12.2].

The assertion **c** is known; e.g., see Section 3.1 of the book [131].

The assertion **d** follows from assertions **b** and **c**. □

8.2.2 Theorem. If A is a right endomorphism-extendable regular ring, then A is a right injective ring.

Proof. Since A is a right endomorphism-extendable ring, A is a right quasicontinuous ring by 8.1.1. By 8.2.1(d) and 1.3.1(e), A is the direct product of a right injective regular ring and a normal ring. Therefore, without loss of generality, we can consider only the case where every principal right ideal of the ring A is generated by a central idempotent. It is sufficient to prove that for an arbitrary right ideal B of the ring A, every homomorphism $f\colon B_A \to A_A$ can be extended to a homomorphism $A_A \to A_A$. Let $b \in B$. Then $bA = eA$ for some central idempotent $e \in bA \subseteq B$. Then $f(b) = f(be) = f(b)e = ef(b) \in B$. Therefore, $f(B) \subseteq B$. Since A_A is an endomorphism-extendable module, the homomorphism $f\colon B_A \to A_A$ can be extended to a homomorphism $A_A \to A_A$. □

8.2.3 Right completely integrally closed subrings and right classically completely integrally closed subrings. Let Q be a ring and A a unitary subring in Q.

A is said to be *right completely integrally closed* subring in Q if A contains any element $q \in Q$ such that $qB \subseteq B$ for some essential right ideal B of the ring A.

A is said to be *right classically completely integrally closed* subring in Q if A contains any element $q \in Q$ such that $q^n a \in A$ for some nonzero divisor a of the ring A and all positive integers n.

Let A be a ring and $Q = Q_{\max}(A)$ the maximal right ring of fractions[1] of the ring A.

a. If the ring Q is right injective, then A is a right completely integrally closed subring in Q if and only if A is a right strongly endomorphism-extendable ring.

b. If A is a right nonsingular ring and Q is the maximal right ring of fractions of the ring A, then A is a right completely integrally closed subring in Q if and only if A is a (strongly) right endomorphism-extendable ring.

1 The definition and main properties of the ring $Q_{\max}(A)$ are given in many books. For example, see [62, Section 16] [151, Section 14.4]

Proof. **a.** Since the ring Q is right injective, it is well known that Q_A is the injective hull of the module A_A and the ring End Q_A can be naturally identified with the ring Q; e.g., see [151, Section 14.4, Proposition 4.1]. Therefore, **b** follows from 8.1.6.

b. Since the ring A is right nonsingular, it is well known that the ring Q is right injective; e.g., see [62, Theorem 16.12]. Therefore, **b** follows from **a** and 8.1.7. □

8.2.4 Proposition. Let A be a right Goldie semiprime ring and let Q be the semisimple Artinian right classical ring of fractions of the ring A (see 7.2.1). The following conditions are equivalent.

1) A is a right classically completely integrally closed subring in Q.

2) A is a right completely integrally closed subring in Q.

3) A is a right strongly endomorphism-extendable ring.

4) A is a right endomorphism-extendable ring.

Proof. By 7.2.1(a), the semiprime right Goldie ring A is right nonsingular. In addition, the semisimple Artinian ring Q is the maximal right ring of fractions of the ring A by [62, Theorem16.14]. Therefore, the equivalence of conditions 2, 3 and 4 follows from 8.2.3(b).

1) $\Rightarrow$ 2). Let B be an essential right ideal of the ring A and let q be an element of the ring Q such that $qB \subseteq B$. By 7.2.1(a), the essential right ideal B contains some nonzero divisor b. Since $qB \subseteq B$, we have that $q^n B \subseteq B$ for all nonnegative integers n. Therefore, $q^n b \in B \subseteq A$ for all nonnegative integers n. Since A is a right classically completely integrally closed subring in Q, we have that $q \in Q$ and A is a right completely integrally closed subring in Q.

2) $\Rightarrow$ 1). Let Q be an element of the ring Q such that $q^n b \in A$ for some nonzero divisor $b \in A$ and all nonnegative integers n. We denote by B the right ideal $\sum_{n=0}^{\infty} q^n bA$ of the ring A. Since B contains the nonzero divisor b, we have that B is an essential right ideal by 7.2.1(a). In addition, $qB \subseteq B$. Since A is a right completely integrally closed subring in Q, we have that $q \in Q$ and A is a right classically completely integrally closed subring in Q. □

8.2.5 Theorem. For a ring A, the following conditions are equivalent.

1) A is a right endomorphism-extendable, right nonsingular ring.

2) A is a right strongly endomorphism-extendable, right nonsingular ring.

3) $A = B \times C$, where B is a right injective regular ring, C is a left invariant, reduced Baer ring and C is a right completely integrally closed subring of its maximal right rings of fractions Q.

Proof. The implication 1) $\Rightarrow$ 2) follows from 8.1.7.

2) $\Rightarrow$ 3). By 8.1.1, every endomorphism-extendable module is quasicontinuous. Therefore, it follows from 8.2.1(d) that $A = B \times C$, where B is a right injective regular ring and C is a right strongly endomorphism-extendable Baer reduced ring. By 8.2.3(b), C is a right completely integrally closed subring in Q. By 3.3.7(f), C is a left invariant ring.

The implication 3) $\Rightarrow$ 1) follows from 8.2.3(b). □

8.2.6 Corollary. For a ring A, the following conditions are equivalent.

1) A is a right endomorphism-extendable, right nonsingular indecomposable ring.

2) A is a right strongly endomorphism-extendable, right nonsingular indecomposable ring.

3) Either A is a right injective regular indecomposable ring or A is a left invariant, right and left Ore domain which is a right classically completely integrally closed subring of its classical division ring of fractions Q.

Proof. The implication 3) $\Rightarrow$ 1) follows from Proposition 8.2.4.

The implication 1) $\Rightarrow$ 2) follows from 8.1.7.

2) $\Rightarrow$ 3). By Theorem 8.2.5, A is either a right injective regular indecomposable ring or a left invariant indecomposable Baer reduced ring which is a right completely integrally closed subring of its maximal right rings of fractions Q. It is sufficient to consider only the second case. The left invariant normal indecomposable Baer ring A is a left invariant domain. By 8.1.2(e), A is a right and left Ore domain. Let Q be the classical division ring of fractions of the domain A. By Proposition 8.2.4, A is a right classically completely integrally closed subring in Q. □

8.2.7 Theorem. Let A be a right endomorphism-extendable ring.

a. If every essential right ideal of A is an ideal, then A is a right strongly endomorphism-extendable ring.

b. If for any essential right ideal N of the ring A, the module A/N is semi-Artinian, then A is a right strongly endomorphism-extendable ring.

c. If for any element $a \in A$ and each element $s \in \operatorname{Sing} A_A$, there exists a positive integer n such that $s^{n+1}a \in \sum_{i=0}^{n} s^i aA$, then A is a right strongly endomorphism-extendable ring.

d. If every element of the ideal $\operatorname{Sing} A_A$ is nilpotent, then A is a right strongly endomorphism-extendable ring.

e. If A is a right Noetherian ring, then A is a right strongly endomorphism-extendable ring.

Theorem 8.2.7 follows from 8.1.8.

8.2.8. Let A be a right strongly endomorphism-extendable ring.

a. $\operatorname{Sing} A_A \subseteq J(A)$.

b. If A has an essential quasi-injective right ideal, then the ring A is right injective. In particular, if A has an essential semisimple right ideal, then the ring A is right injective.

c. If $\operatorname{Sing} A_A$ is an essential right ideal of the ring A, then the ring A is right injective.

8.2.8 follows from Proposition 8.1.9(a)–8.1.9(c) and 6.1.1(e).

8.2.9 Theorem. For a ring A, the following conditions are equivalent.

1) A is a right strongly endomorphism-extendable ring without nontrivial idempotents.

2) Either A is a right injective, right uniform local ring and $J(A) = \operatorname{Sing} A_A$ is a nonzero ideal or A is a left invariant, right and left Ore domain that is a right classically completely integrally closed subring of its classical division ring of fractions.

Proof. The implication 2) ⇒ 1) follows from Corollary 8.2.6.

1) ⇒ 2). Since A is a ring without of nontrivial idempotents, A_A is an indecomposable module. By Proposition 8.1.9(d), A is a right uniform ring and either A is right quasi-injective and $A_A = G(A_A)$ or A is a domain. By Corollary 8.2.6, it is sufficient to consider the case, where the ring A is right quasi-injective. By 6.1.1(e), A is a right injective ring. Then $J(A) = \operatorname{Sing} A_A$ [63, Theorem 19.27]. Since A is a right injective ring without nontrivial idempotents, A is a local ring. □

8.3 Rings with endomorphism-extendable cyclic modules

8.3.1. For a module M, the following conditions are equivalent.

1) M is an endomorphism-liftable module and each of its factor modules is endomorphism-extendable.

2) M is an endomorphism-extendable module and each of its submodules is endomorphism-liftable.

Proof. 1) ⇒ 2). Let N be a submodule in M, $\overline{f}$ be an endomorphism of the the factor module $\overline{N} = N/P$, and let $h\colon M \to M/P$ be the natural epimorphism. Since M/P is an endomorphism-extendable module, $\overline{f}$ can be extended to an endomorphism $\overline{g}$ of the module M/P. Since M is an endomorphism-liftable module, $\overline{g}h = hg$ for some $g \in \operatorname{End} M$. Therefore, $g(N) \subseteq N$. Then g induces $f \in \operatorname{End}(N)$ and $\overline{f}h_N = h_N f$, where $h_N\colon N \to N/P$ is the natural epimorphism. Therefore, N is an endomorphism-liftable module.

2) ⇒ 1). Let M/P be an arbitrary factor module, $h\colon M \to M/P$ be the natural epimorphism, N/P be a submodule in M/P, h_N be the restriction of h to N, and let $\bar{f}$ be an endomorphism of the module N/P. By assumption, N is an endomorphism-liftable module. Therefore, there exists an endomorphism f of the module N with $\bar{f}h_N = h_N f$. Since M is an endomorphism-extendable module, the endomorphism f can be extended to an endomorphism g of the module M/P and $g(P) = f(P) \subseteq P$. Since $g(P \subseteq P$, we have that g induces the endomorphism $\bar{g}$ of the module M/P, which is an extension of the endomorphism $\bar{f}$ of the module N/P. Therefore, M/P is an endomorphism-extendable module. □

8.3.2 Rings with quasicontinuous cyclic modules. Let A be a ring such that all cyclic right A-modules are quasicontinuous. The following properties are familiar, e.g., see [167] or [57, 14.7].
$A = B \times C$, where B is a semisimple Artinian ring, $C = C_1 \times \cdots \times C_n$ and all C_i are right uniform rings;
The ring C_i is local if and only if C_i is a right uniserial ring.
If all cyclic right A-modules are quasi-injective, then for any i it is true that C_i is a right injective, right and left uniserial, right and left invariant ring and $J(C_i)$ is a nil ideal.

8.3.3 Proposition. For a ring A, the following conditions are equivalent.

1) Every cyclic right A-module is endomorphism-extendable.

2) A is a right endomorphism-extendable ring and every right ideal of the ring A is endomorphism-liftable.

3) $A = B \times C$, where B is a semisimple Artinian ring, $C = C_1 \times \cdots \times C_n$ and all C_i are right uniform rings, in which all right ideals are endomorphism-liftable.

4) $A = B \times C$, where B is a semisimple Artinian ring, $C = C_1 \times \cdots \times C_n$ and all C_i are right uniform rings such that all cyclic A-modules are endomorphism-extendable.

Proof. The projective module A_A is an endomorphism-liftable. Therefore, the equivalences 1) ⇔ 2) and 3) ⇔ 4) follow from 8.3.1.

The implication 4) ⇒ 1) is directly verified.

1) ⇒ 4). Since every endomorphism-extendable module is quasicontinuous, all cyclic right A-modules are quasicontinuous. By 8.3.2, $A = B \times C$, where B is a semisimple Artinian ring, $C = C_1 \times \cdots \times C_n$ and all C_i are right uniform rings. It is clear that all cyclic right C_i-modules are endomorphism-extendable. □

8.3.4 Proposition. For a ring A, the following conditions are equivalent.

1) A is a right strongly endomorphism-extendable ring and every cyclic right A-module is endomorphism-extendable.

2) A is a right strongly endomorphism-extendable ring and every right ideal of the ring A is endomorphism-liftable.

3) $A = A_1 \times \cdots \times A_n$ and for every A_i, each right ideal of the ring A_i is endomorphism-liftable and either A_i is a simple Artinian ring or A_i is a right uniserial, right injective ring, and $J(A)$ coincides with nonzero ideal $\operatorname{Sing} A_A$ or A_i is a left invariant, right and left Ore domain which is right classically completely integrally closed subring of its classical division ring of fractions.

Proof. The equivalence 1) ⇔ 2) follows from 8.3.3.

The implication 3) ⇒ 2) follows from Corollary 8.2.6.

2) ⇒ 3). By Proposition 8.3.3, it is sufficient to consider the case where A is a right uniform ring.

If A is a domain, it follows from Corollary 8.2.6 that A is a left invariant, right and left Ore domain, which is a right classically completely integrally closed subring of its classical division ring of fractions.

We assume that A is not a domain. Since A is a right uniform ring with left zero divisors, $\operatorname{Sing} A_A$ is an essential right ideal. By Theorem 8.2.9, A is a right injective, right uniform local ring and $J(A) = \operatorname{Sing} A_A$ is a nonzero ideal. By 8.3.2, A is a right uniserial ring. □

8.3.5 Rings that are integral over their centers. A ring A is said to be *integral over its center* if for any element $s \in A$, there exist central elements $c_1, \ldots, c_n$ of the ring A such that $s^{n+1} = \sum_{i=0}^{n} s^i c_i$.

If a ring A is integral over its center, then the following conditions are equivalent.

1) A is a right strongly endomorphism-extendable ring and every cyclic right A-module is endomorphism-extendable.

2) A is a right strongly endomorphism-extendable ring and every right ideal of the ring A is endomorphism-liftable.

3) $A = A_1 \times \cdots \times A_n$ and for every A_i,each right ideal of the ring A_i is endomorphism-liftable and either A_i is a simple Artinian ring or A_i is a right uniserial, right injective ring and $J(A)$ coincides with the nonzero ideal $\operatorname{Sing} A_A$ or A_i is a left invariant, right and left Ore domain which is right classically completely integrally closed subring of its classical division ring of fractions.

8.3.5 follows from Proposition 8.3.4 and Theorem 8.2.7(c).

8.3.6 Theorem. If all essential right ideals of a ring A are ideals, then the following conditions are equivalent.

1) Every cyclic right A-module is endomorphism-extendable.

2) $A = A_1 \times \cdots \times A_n$ and for every A_i, we have that every right ideal of the ring A_i is endomorphism-liftable and either A_i is a simple Artinian ring or A_i is a right invariant, right uniserial, right injective ring and $J(A)$ coincides with the nonzero ideal $\operatorname{Sing} A_A$, or A_i is an invariant domain, which is a right classically completely integrally closed subring of its classical division ring of fractions.

Proof. The implication 2) $\Rightarrow$ 1) follows from Proposition 8.3.4.

1) $\Rightarrow$ 2). Proposition 8.3.3 argues the case where A is a right uniform ring. Then all nonzero right ideals of the ring A are essential. Since all essential right ideals of the ring A are ideals, the right uniform ring A is right invariant. By Theorem 8.2.7(a), A is a right strongly endomorphism-extendable ring. Now we use Proposition 8.3.4. □

We need the following well known projectivity criterion.

8.3.7 Dual basis lemma. For a module M, the following conditions are equivalent.

1) M is a projective module.

2) There exist a subset $\{m_i\}_{i\in I} \subseteq M$ and a set $\{f_i\}_{i\in I}$ of homomorphisms $f_i \colon M \to A_A$ such that $m = \sum_{i\in I} m_i f_i(m)$ for any $m \in M$, where $f_i(m) = 0$ for almost of all subscripts i.

3) There exist a system $\{m_i\}_{i\in I}$ of generators of the module M and a set $\{f_i\}_{i\in I}$ of homomorphisms $f_i \colon M \to A_A$ such that $m = \sum_{i\in I} m_i f_i(m)$ for any $m \in M$, where $f_i(m) = 0$ for almost all subscripts i.

Proof. 1) $\Rightarrow$ 2). By 6.1.1(j), we can assume that $M \oplus P = Q_A$, where Q_A is a free module with basis $\{x_i\}_{i\in I}$. Let $g_i \colon x_i A \to A_A$ be isomorphisms such that $g(x_i) = 1$, $t \colon Q \to M$ be the projection with kernel P, $m_i \equiv t(x_i)$, $h_i \colon Q \to x_i A$ be a natural projections, and let $f_i \equiv g_i h_i|_M \colon M \to A_A$. We consider $m \in M$. There exists a finite subset $J \subseteq I$ such that $m = \sum_{i\in J} x_i a_i$. Since

$$\sum_{i\in J} m_i f_i(m) = \sum_{i\in J} m_i g_i(h_i(m)) = \sum_{i\in J} m_i g_i(x_i a_i) = \sum_{i\in J} m_i a_i = m\,,$$

the sets $\{m_i\}_{i\in I}$ and $\{f_i\}_{i\in I}$ have the required properties.

2) $\Rightarrow$ 3). Since $m = \sum_{i\in I} m_i f_i(m)$ for all $m \in M$, the set $\{m_i\}_{i\in I}$ generates M.

3) $\Rightarrow$ 1). Let Q_A be a free module with basis $\{x_i\}_{i\in I}$, $u_i \colon A_A \to x_i A$ be isomorphisms such that $u_i(1) = x_i$, and let $t \colon Q \to M$ be an epimorphism such that $t(x_i) = m_i$ for any $i \in I$. We define a homomorphism $f \colon M \to Q$ by the rule $f(m) = \sum_{i\in I} u_i(f_i(m)) = \sum_{i\in I} x_i f_i(m)$. It can be verified that f is well-defined. Then

$$(tf)(m) = t\left(\sum_{i\in I} x_i f_i(m)\right) = \sum_{i\in I} m_i f_i(m) = m\,.$$

Therefore, $tf = 1_M$ and the module M is isomorphic to a direct summand of the free module Q. By 6.1.1(j), M is a projective module. □

8.3.8. Let A be a unitary subring of the ring B, M be a submodule of the module B_A, and let there exist $m_1, \dots, m_n \in M$ and $b_1, \dots, b_n \in B$ such that $1 = \sum_{i=1}^n m_i b_i$ and $b_i M \subseteq A$ for all i. Then $M = \sum_{i=1}^n m_i A$ is a projective n-generated module.

Proof. Let $f_1, \dots, f_n \colon M_A \to A_A$ be homomorphisms such that $f(m) = b_i m$ for $m \in M$. Then $m = \sum_{i=1}^n m_i f_i(m)$ for any $m \in M$ and the assertion follows from 8.3.7. □

8.3.9 ([149]). If A is a left semihereditary ring and for any positive integer n and the ring of all of $n \times n$ matrices does not contain an infinite set of orthogonal idempotents, then A is a right semihereditary ring.

8.3.10 Proposition. Let A be a right endomorphism-extendable ring such that all its right ideals are π-projective.

a. In the ring A, for any right ideals B and C, there exist two elements $x, y \in A$ such that

$$x(B+C) \subseteq B\,, \quad y(B+C) \subseteq C\,, \quad xB + yC \subseteq B \cap C\,,$$
$$(x+y-1)(B+C) \subseteq B \cap C\,,\ B + C = (x+y)(B+C) + B \cap C\,,$$
$$(B+C) \cap r(x+y) \subseteq B \cap C\,.$$

b. For any two right ideals B and C of the ring A, there exist elements $s, t \in A$ such that $s + t = 1$ and $sB + tC \subseteq B \cap C$.

c. If $d_1, \dots, d_n$ are nonzero divisors of the ring A and $D = \sum_{i=1}^{n} Ad_i$, then the finitely generated left ideal D is a projective left A-module.

d. If A is a domain, then A is a left invariant, right and left semihereditary, right and left Ore domain.

Proof. **a.** We set $M = B + C$. Since M_A is a π-projective module, it follows from 8.3.1(b) that there exist homomorphisms $f\colon M \to B$ and $g\colon M \to C$ such that

$$f(B) + g(C) \subseteq B \cap C\,, \quad (f + g - 1_M)(M) \subseteq B \cap C\,,$$
$$M = (f+g)(M) + B \cap C\,, \quad \operatorname{Ker}(f+g) \subseteq B \cap C\,.$$

Since f, g are endomorphisms of the right ideal M of the right endomorphism-extendable ring A, there exist elements $x, y \in A$ such that $f(m) = xm \in B$ and $g(m) = ym \in C$ for all $m \in M$. Therefore,

$$xM \subseteq B\,, \quad yM \subseteq C\,, \quad xB + yC \subseteq B \cap C\,,$$
$$(x+y-1)M \subseteq B \cap C\,, \quad M = (x+y)M + B \cap C\,,$$
$$M \cap r(x+y) \subseteq B \cap C\,.$$

b. By **a**, there exist elements $x, y \in A$ such that $xB + yC \subseteq B \cap C$ and $(x+y-1)(B+C) \subseteq B \cap C$. We set $s = y$, $t = 1 - y = x - (x+y-1)$. Then

$$s + t = 1\,, \quad sB + tC \subseteq yB + xC + (x+y-1)C \subseteq B \cap C\,.$$

c. By 3.3.7(g), A has the left classical ring of fractions Q and $d_iAd_i^{-1} \subseteq A$ for any nonzero divisor d_i. Therefore, $d_1 \cdot \dots \cdot d_n d_i^{-1} \equiv a_i \in A$ for all i. All a_i are nonzero divisors in A. Therefore, $a_i^{-1} \in Q$. By **a**, for the right ideals $a_{n-1}A$ and a_nA, there exists an element $t_{n-1} \in A$ such that

$$(1 - t_{n-1})a_{n-1}A + t_{n-1}a_nA \equiv B_{n-1} \subseteq a_{n-1}A \cap a_nA\,.$$

By **a**, for the right ideals $a_{n-2}A$ and B_{n-1}, there exists an element $t_{n-2} \in A$ such that

$$(1 - t_{n-2})a_{n-2}A + t_{n-2}B_{n-1} \equiv B_{n-2} \subseteq a_{n-2}A \cap B_{n-1} \subseteq a_{n-2}A \cap a_{n-1}A \cap a_nA\,.$$

We assume that $k < n - 1$ and we have the right ideal $B_{n-k} \subseteq \cap_{i=0}^{k} a_{n-i}A$. By **a**, for the right ideals $a_{n-k-1}A$ and B_{n-k} there exists an element $t_{n-k-1} \in A$ such that

$$(1 - t_{n-k-1})a_{n-k-1}A + t_{n-k-1}B_{n-k} \equiv B_{n-k-1} \subseteq a_{n-k-1}A \cap B_{n-k} \subseteq \cap_{i=0}^{k} a_{n-i}A\,.$$

Finally, there exists an element $t_1 \in A$ such that

$$\begin{aligned}
B_1 &= (1 - t_1)a_1A + t_1B_2 \subseteq a_1A \cap B_2 \subseteq \cap_{i=0}^{n-1} a_{n-i}A = \cap_{i=1}^{n} a_iA\,,\\
B_1 &= (1 - t_1)a_1A + t_1((1 - t_2)a_2A + t_2((1 - t_3)a_3A + \ldots t_{n-1}a_nA)\ldots)\\
&= (1 - t_1)a_1A + t_1(1 - t_2)a_2A + t_1t_2(1 - t_3)a_3A + \ldots\\
&\quad + t_1t_2\ldots t_{n-2}(1 - t_{n-1})a_nA + t_1t_2\ldots t_{n-1}a_nA \subseteq \cap_{i=1}^{n} a_iA\,,
\end{aligned}$$

$$(1 - t_1)a_1 \in \cap_{i=1}^{n} a_iA\,, \quad t_1(1 - t_2)a_2 \in \cap_{i=1}^{n} a_iA\,,$$
$$t_1t_2(1 - t_3)a_3 \in \cap_{i=1}^{n} a_iA\,, \quad \ldots\,,$$
$$t_1t_2\ldots t_{n-1}a_n \in \cap_{i=1}^{n} a_iA\,.$$

Let

$$(1 - t_1)a_1 = a_if_{1i} = b_1\,, \quad t_1(1 - t_2)a_2 = a_if_{2i} = b_2\,, \quad \ldots\,,$$
$$t_1t_2\ldots t_{n-1}a_n = a_if_{ni} = b_n\,, \quad M \equiv \sum_{i=1}^{n} Aa_i^{-1} \subseteq Q\,.$$

Then $\sum_{i=1}^{n} b_ia_i^{-1} = 1$. In addition, for any b_j, we have

$$Mb_j = \sum_{i=1}^{n} Aa_i^{-1}b_j = \sum_{i=1}^{n} Aa_i^{-1}a_if_{ji} = \sum_{i=1}^{n} Af_{ji} \subseteq A \subseteq M\,.$$

By 8.3.8, the module ${}_AM$ is projective. Therefore,

$$\begin{aligned}
{}_AD &= \sum_{i=1}^{n} Ad_i(d_1\cdot\cdots\cdot d_n)^{-1}d_1\cdot\cdots\cdot d_n = \sum_{i=1}^{n} Aa_i^{-1}d_1\cdot\cdots\cdot d_n =\\
&= {}_AMd_1\cdot\cdots\cdot d_n \cong {}_AM\,,
\end{aligned}$$

whence ${}_AD$ is a projective module.

d. By **c**, A is a left semihereditary domain. By Corollary 8.2.6, A is a left invariant, right and left Ore domain with a classical division ring of fractions Q. For any positive integer n, the ring Q_n of all $n \times n$ matrices over the division ring Q does not contain an infinite set of orthogonal idempotents. Therefore, its subring A_n does not contain an infinite set of orthogonal idempotents. By 8.3.9, A is a right semihereditary domain. □

8.3.11. Let A be the direct product of Artinian uniserial rings. It is well known and is directly verified that A is an invariant ring such that every cyclic A-module is quasi-injective.

8.3.12 Theorem. For a ring A, the following conditions are equivalent.

1) A is a right Noetherian ring such that all cyclic right A-modules are endomorphism-extendable.

2) A is a left Noetherian ring such that all cyclic left A-modules are endomorphism-extendable.

3) $A = A_1 \times \cdots \times A_n$, where A_i is either a simple Artinian ring or a uniserial Artinian ring, or an invariant hereditary Noetherian domain, $i = 1, \ldots, n$.

Proof. It is sufficient to prove the equivalence of conditions 1) and 3).

1) $\Rightarrow$ 3). By Theorem 8.2.7(e), A is a right strongly endomorphism-extendable ring. By Proposition 8.3.4, it is sufficient to consider the case, where either A is a right uniserial, right injective ring or A is a left invariant domain.

We assume that A is a right uniserial, right injective ring. Since A is a right Noetherian right injective ring, A is an Artinian ring [63, Theorem 24.5]. By Theorem 6.1.10, all cyclic right A-modules are quasi-injective. By 8.3.2, A is both right and left uniserial, and right and left invariant ring.

We assume that A is a left invariant domain. Since A is a right Noetherian, left invariant ring, we have that A is a left Noetherian ring. By Proposition 8.3.10(d), A is a right and left semihereditary domain. The right and left Noetherian, right and left semihereditary domain A is a hereditary domain. By 7.2.3(c), A is an invariant hereditary Noetherian domain.

3) $\Rightarrow$ 1). By Proposition 8.3.3, we can assume that A is either a uniserial Artinian ring or an invariant hereditary Noetherian domain. Let M be a nonzero cyclic right A-module. If A is a uniserial Artinian ring, then M is an endomorphism-extendable module by 8.3.11.

We assume that A is an invariant hereditary Noetherian domain. If M is a nonsingular module, then M is a uniform module and M is an endomorphism-extendable module by Theorem 7.2.11. If the cyclic module M is not a nonsingular, then M has nonzero annihilator B. By 7.2.3(c), A/B is a finite direct product of invariant uniserial Artinian rings. By 8.3.11, M is a quasi-injective A/B-module. Therefore, M is a quasi-injective A-module. In particular, M is an endomorphism-extendable module. □

8.3.13 The completion of the proof of Theorems 8A and 8B. Theorems 8A and 8B follow from Theorems 8.2.5 and 8.3.12, respectively.

8.3.14 Open question. Let A be a ring such that every cyclic right A-module is endomorphism-extendable. Is it true that every cyclic left A-module is endomorphism-extendable?

9 Automorphism-invariant modules and rings

The main results of this section are Theorems 9A, 9B, 9C and 9D.

9A Theorem (Tuganbaev [185]). Let A be a right strongly semiprime ring. If X is a right A-module and there exists an essential right ideal B of the ring A such that X is injective with respect to the module B_A, then X is an injective module.

9B Theorem (Tuganbaev [189]). If A is a ring with right Goldie radical G, then the following conditions are equivalent.
1) Every nonsingular right A-module X, which is injective with respect to some essential right ideal of the ring A, is an injective module.
2) A/G is a right strongly semiprime ring.

9C Theorem (Tuganbaev [189]). For a ring A with right Goldie radical G, the following conditions are equivalent.
1) A/G is a semiprime right Goldie ring.
2) Any direct sum of automorphism-invariant nonsingular right A-modules is an automorphism-invariant module.
3) Any direct sum of automorphism-invariant nonsingular right A-modules is an injective module.

9D Theorem (Tuganbaev [186]). A ring A is a right automorphism-invariant, right nonsingular ring if and only if $A = S \times T$, where S is a right injective regular ring and T is a strongly regular ring that contains all invertible elements of its maximal right ring of fractions.

Remark. The completion of the proof of Theorems 9A–9D is given in 9.2.6.

9.1 Automorphism-invariant modules

9.1.1. Let A be a ring and Y a nonsingular right A-module. If $\{y_i\}_{i\in Y}$, $|I| \geq 2$, is a subset of the module Y such that $\mathrm{Hom}(Y_i, Y_j) = 0$ for any submodules $Y_i \subseteq y_iA$ and $Y_j \subseteq y_jA$ for all $i \neq j$, then there exists a set $\{B_i\}_{i\in I}$ of nonzero right ideals of the ring A such that for any i, the right A-module B_i is isomorphic to a submodule in y_iA and $B_iB_j = 0$ for all $i \neq j$.

Proof. In I, we fix distinct subscripts i and j. By 6.2.3(f), there exist nonzero right ideals B_i and B_j of the ring A such that y_iA contains a nonzero submodule Y_i that is isomorphic to the right A-module B_i, and y_jA contains a nonzero submodule Y_j that is isomorphic to the right A-module B_j. By assumption, $\mathrm{Hom}(Y_i, Y_j) = 0$. Then $\mathrm{Hom}(B_i, B_j) = 0$, since $B_i \cong Y_i$ and $B_j \cong Y_j$. For any element $b_j \in B_j$, a homomorphism from B_i into B_j is

https://doi.org/10.1515/9783110659825-009

defined by the rule $x \to b_j x$, $x \in B_i$. Since $\mathrm{Hom}(B_i, B_j) = 0$, we have $b_j B_i = 0$, whence $B_i B_j = 0$. □

9.1.2. Let A be a ring and let M be a nonsingular automorphism-invariant A-module.

a. [60, Theorem 3, Theorem 6]. There exists a direct decomposition $M = X \oplus Y$ such that X is a quasi-injective nonsingular module, Y is a square-free[1] nonsingular automorphism-invariant module, the modules X, Y are injective with respect to each other, any sum of closed submodules of the module Y is an automorphism-invariant module, $\mathrm{Hom}(X, Y) = \mathrm{Hom}(Y, X) = 0$ and $\mathrm{Hom}(Y_1, Y_2) = 0$ for any submodules Y_1 and Y_2 in Y with $Y_1 \cap Y_2 = 0$.

b. If Y is a square-free direct summand of the module M and Y is an essential extension of direct sum of uniform modules, then Y is an essential extension of a quasi-injective module Z that is the direct sum of uniform quasi-injective modules.

c. If Y is a finite-dimensional square-free direct summand of the module M, then Y is an essential extension of a quasi-injective module, which is the finite, direct sum of uniform quasi-injective modules.

d. If Y is a square-free direct summand of the module M which is not finite-dimensional, then there exists an infinite set $\{B_i\}_{i=1}^{\infty}$ of nonzero right ideals of the ring A such that $B_i B_j = 0$ for all $i \neq j$.

e. If A is a finite subdirect product of prime rings and Y is a square-free direct summand of the module M from **a**, then Y is a finite-dimensional module which is an essential extension of quasi-injective modules $Y_1 \oplus \cdots \oplus Y_n$, where all Y_i are quasi-injective uniform modules.

f. If the ring A is right strongly semiprime and Y is a square-free direct summand of the module M from **a**, then Y is an injective module.

g. If the factor ring $A/G(A_A)$ is right strongly semiprime and Y is a square-free direct summand of the module M from **a**, then Y is an injective module.

Proof. **a.** The assertion has been proven in [60, Theorem 3, Theorem 6].

b. By assumption, Y is an essential extension of direct sum of uniform modules Y_i, $i \in I$. For every i, the uniform module Y_i is an essential submodule of some closed uniform submodule Z_i of the module Y. We set $Z = \sum_{i\in I} Z_i$. Then $Z = \oplus_{i\in I} Z_i$ and Y is an essential extension of the module Z. By **a**, Z is an automorphism-invariant module. In addition, Z is the direct sum of uniform modules. By 6.1.13(b), the module Z is quasi-injective. All uniform direct summands Z_i of the quasi-injective module Z are quasi-injective.

c. The finite-dimensional module Y is an essential extension of the finite direct sum of uniform modules. By **b**, Y is an essential extension of a quasi-injective module that is the finite direct sum of uniform quasi-injective modules.

---

1 A module M is said to be *square-free* if M does not have nonzero submodules $X \oplus Y$ such that $X \cong Y$.

d. Since the module Y is not finite-dimensional, Y contains the infinite direct sum $\oplus_{i=1}^{\infty} y_i A$ of nonzero cyclic submodules. For any distinct positive integers i, j and arbitrary submodules $Y_i \subseteq y_i A$, $Y_j \subseteq y_j A$, we have $Y_i \cap Y_j \subseteq y_i A \cap y_j A = 0$, whence $\mathrm{Hom}(Y_i, Y_j) = 0$ by **a**. By 9.1.1, there exists a set $\{B_i\}_{i=1}^{\infty}$ of nonzero right ideals of the ring A such that for any positive integer i, the right A-module B_i is isomorphic to a submodule in $y_i A$ and $B_i B_j = 0$ for any $i \neq j$.

e. By **c**, it is sufficient to prove that Y is a finite-dimensional module. We assume the contrary. By **d**, there exists an infinite set $\{B_i\}_{i=1}^{\infty}$ of nonzero right ideals of the ring A such that $B_i B_j = 0$ for all $i \neq j$. Since A is a finite subdirect product of prime rings, there exists a finite set $\{P_k\} \sum_{k=1}^{n}$ prime ideals of the ring A such that $P_1 \cap \cdots \cap P_n = 0$. For all i, we have $B_i \neq 0$; in addition, $P_1 \cap \cdots \cap P_n = 0$. Therefore, for any positive integer i, there exists a prime ideal $P_{\alpha(i)} \in \{P_i\}_{i=1}^{n}$ such that B_i is not contained in $P_{\alpha(i)}$. Since $P_{\alpha(i)}$ is a prime ideal and $B_i B_j = 0 \subseteq P_{\alpha(i)}$ for all $j \neq i$, we have that B_j is contained in $P_{\alpha(i)}$ for all $j \neq i$. In addition, B_j is not contained in $P_{\alpha(j)}$. This implies that all ideals $P_{\alpha(i)}$ are distinct. This contradicts the property that $\{P_k\} \sum_{k=1}^{n}$ is a finite set.

f. Since the ring A is right strongly semiprime, A is the finite subdirect product of prime rings, see [86, Theorem 1]. By **e**, Y is an essential extension of some quasi-injective of the module Y'. By Theorem 7.2.17, all nonsingular quasi-injective right A-modules are injective. Therefore, Y' is an injective essential submodule of the module Y. Then $Y = Y'$ and the module Y is injective.

g. Since the module Y is nonsingular, $G(Y) = 0$. Then $YG(A_A) \subseteq G(M) = 0$. Therefore, Y is a natural right $A/G(A_A)$-module. It is directly verified that Y is a nonsingular square-free $A/G(A_A)$-module. With the use of 6.1.3, it is verified that Y is an automorphism-invariant $A/G(A_A)$-module. Since the factor ring $A/G(A_A)$ is right strongly semiprime, it follows from **e** that Y is an injective $A/G(A_A)$-module. Therefore, Y is a quasi-injective A-module. By Theorem 7.2.17, Y is an injective A-module. □

9.1.3 Theorem. Let A be a ring with right Goldie radical $G(A_A)$. The following conditions are equivalent.

1) All automorphism-invariant nonsingular right A-modules are injective.

2) $A/G(A_A)$ is a right strongly semiprime ring.

Proof. The implication 1) ⇒ 2) follows from Theorem 7.2.17 and the property that every quasi-injective module is an automorphism-invariant.

2) ⇒ 1). Let M be a automorphism-invariant A–nonsingular right module. By 9.1.2(a), there exists a direct decomposition $M = X \oplus Y$ such that X is a quasi-injective module, Y is a square-free nonsingular automorphism-invariant module. By 9.1.2(g), Y is an injective module. By Theorem 7.2.17, the module X is injective. Then $M = X \oplus Y$ is an injective module. □

9.1.4 Remark. A nonsingular automorphism-invariant module is not necessarily quasi-injective; see the example VIII(c) from the introduction.

9.1.5. Let A be a ring, X be a nonsingular nonzero right A-module, $\{C_i \mid i \in I\}$ be the set of all of nonzero right ideals of the ring A such that every nonzero submodule A-module C_i is not isomorphic to a submodule of the module X, and let $\{D_j \mid j \in J\}$ be the set of all nonzero right ideals D_j of the ring A such that D_j is isomorphic to submodule of the module X. We set $C = \sum_{i \in I} C_i$, $D = \sum_{j \in J} D_j$ and $B = C + D$.

a. For any submodule C' of the module C_A, every homomorphism $f: C'_A \to X$ is the zero homomorphism.

b. The module X is injective with respect to the module C_A.

c. B is an essential right ideal of the ring A.

d. If the module X is quasi-injective, then X is injective with respect to the essential right ideal B.

Proof. **a.** We assume that $f \neq 0$. Since X is a nonsingular module and $C'/\operatorname{Ker} f \cong f(C') \subseteq X$, we have that $\operatorname{Ker} f$ is not an essential submodule in C'_A. There exists a nonzero element $c \in C'$ with $cA \cap \operatorname{Ker} f = 0$. The nonzero submodule cA of the module C' is isomorphic to the nonzero submodule $f(cA)$ of the module X. Therefore, $f(c) \neq 0$. There exists a finite subset K in I such that $c = \sum_{k \in K} c_k$ and $c_k \in C_k$ for all $k \in K$. Since $f(c) \neq 0$, we have that $f(c_k) \neq 0$ for some $k \in K \subseteq I$. Therefore, $c_k A$ is a nonzero submodule A-module C_k that is isomorphic to a nonzero submodule of the module X. This contradicts the property that $C_k \in \{C_i \mid i \in I\}$.

b. The assertion follows from **a.**

c. We assume that B is not an essential right ideal. Then $B \cap E = 0$ for some nonzero right ideal E. Then $C \cap E = 0$ and $D \cap E = 0$. Since $C \cap E = 0$, we have $E \notin \{C_i \mid i \in I\}$. Therefore, there exists a nonzero submodule E_1 of the module E that is isomorphic to a submodule of the module X. Then $E_1 \in \{D_j \mid j \in J\}$. Therefore, $E_1 \subseteq D \cap E = 0$. This is a contradiction.

d. Since X is a quasi-injective module, X is injective with respect to any module which is isomorphic to a submodule of the module X. Therefore, X is injective with respect to each the A-module D_j. By 6.1.1(f), the module X is injective with respect to the module D_A. In addition, X is injective with respect to the module C_A by **b.** By 6.1.1(f), the module X is injective with respect to the module $C + D = B$. □

9.1.6 Theorem. Let A be a right strongly semiprime ring and X a right A-module. If there exists an essential right ideal B of the ring A such that X is injective with respect to the module B_A, then X is an injective module.

Proof. By 7.1.3(a), X is injective with respect to the module $(AB)_A$, where AB is an ideal generated by right ideal B. Since B is an essential right ideal and $B \subseteq AB$, the ideal

AB is an essential right ideal. Since A is a right strongly semiprime ring, the ideal AB contains a finite subset $K = \{k_1, \ldots, k_n$ with the zero right annihilator $r(K)$. Since $r(K) = r(k_1) \cap \cdots \cap r(k_n) = 0$, the module A_A is isomorphic to a submodule of the direct sum of n of copies of the module $(AB)_A$. In addition, the module X is injective with respect to the module $(AB)_A$. By 6.1.1(d), the module X is injective. □

9.1.7 Example. In connection to Theorem 9.1.6, we remark that there exist a finite commutative ring A, an essential ideal B of the ring A and a noninjective B-injective A-module X. We denote by A, B and X the finite commutative ring $\mathbb{Z}/4\mathbb{Z}$, the ideal $2\mathbb{Z}/4\mathbb{Z}$, and the module B_A, respectively. Then B is an essential ideal and the module X is injective with respect to B_A. Since X is not a direct summand in A_A, we have that X is not an injective module.

9.1.8 Corollary. For a ring A, the following conditions are equivalent.

1) A is a right strongly semiprime ring.

2) Every right A-module, which is injective with respect to some essential right ideal of the ring A, is an injective module and A is right nonsingular.

Corollary 9.1.8 follows from Theorem 9.1.6 and the property that every right strongly semiprime ring is right nonsingular [86].

9.1.9. Let A be a ring, $G = G(A_A)$ be the right Goldie radical of the ring A, $h\colon A \to A/G$ be the natural ring epimorphism, and let X be a nonsingular right A-module.

a. If B is an essential right ideal of the ring A, then $h(B)$ is an essential right ideal of the ring $h(A)$.

b. If B is a right ideal of the ring A such that $G \subseteq B$ and $h(B)$ is an essential right ideal of the ring $h(A)$, then B is an essential right ideal of the ring A.

c. $MG \subseteq G(M)$ for any right A-module M.

d. $XG = 0$ and a natural $h(A)$-module X is nonsingular. In addition, if Y is an arbitrary nonsingular right A-module, then $YG = 0$ and the $h(A)$-module homomorphisms $Y \to X$ coincide with the A-module homomorphisms $Y \to X$. Consequently, the A-module X is Y-injective if and only if the $h(A)$-module X is Y-injective. The essential submodules of the $h(A)$-module X coincide with the essential submodules of the A-module X.

e. X is an injective $h(A)$-module if and only if X is an injective A-module.

f. $X_{h(A)}$ is an essential extension of direct sum of uniform modules if and only if X_A is an essential extension of direct sum of uniform modules.

g. X_A is an essential extension of some module, which is the direct sum of submodules such that each of the submodules is isomorphic to some nonzero right ideal of the ring A.

h. If the ring A is right finite-dimensional, then X_A is an essential extension of some module which is the direct sum of submodules such that each of the submodules is isomorphic to some nonzero uniform right ideal of the ring A.

i. If the ring $h(A)$ is right finite-dimensional, then $X_{h(A)}$ is an essential extension of some module, which is the direct sum of submodules such that each of the submodules is isomorphic to some nonzero uniform right ideal of the ring $h(A)$.

Proof. **a.** We assume that $h(B)$ is not an essential right ideal of the ring $h(A)$. Then there exists a right ideal C of the ring A such that C properly contains G and $h(B) \cap h(C) = h(0)$. Since $h(B) \cap h(C) = h(0)$, we have $B \cap C \subseteq G$. Since C properly contains the closed right ideal G, we have that C_A contains a nonzero submodule D such that $D \cap G = 0$. Since B is an essential right ideal, $B \cap D \neq 0$ and $(B \cap D) \cap G = 0$. Then $h(0) \neq h(B \cap D) \subseteq h(B) \cap h(C) = h(0)$. This is a contradiction.

b. We assume that B is not an essential right ideal of the ring A. Then $B \cap C = 0$ for some nonzero right ideal C of the ring A and $G \cap C \subseteq B \cap C = 0$. Therefore, $h(C) \neq h(0)$. Since $h(B)$ is an essential right ideal of the ring $h(A)$, we have that $h(B) \cap h(C) \neq h(0)$. Let $h(0) \neq h(b) = h(c) \in h(B) \cap h(C)$, where $b \in B$ and $c \in C$. Then $c - b \in G \subseteq B$. Therefore, $c \in B \cap C = 0$, whence $h(c) = h(0)$. This is a contradiction.

c. For any element $m \in M$, the module mG_A is a Goldie radical module, since mG_A is a homomorphic image of the Goldie radical module G. Therefore, $mG \subseteq G(M)$ and $MG \subseteq G(M)$.

d. By **c**, $XG = 0$. We assume that $x \in X$ and $xh(B) = 0$ for some essential right ideal $h(B)$, where $B = h^{-1}(h(B))$ is the complete pre-image of $h(B)$ in the ring A. By **b**, B is an essential right ideal of the ring A. Then $xB = 0$ and $x \in \operatorname{Sing} X = 0$. Therefore, X is a nonsingular $h(A)$-module. The remaining part **d** is directly verified.

e. Let R be one of the rings A, $h(A)$ and let M be a right R-module. By 6.1.1(e), the module M is injective if and only if M is injective with respect to the module R_R. Now the assertion follows from 4.

f. The assertion follows from **d**.

g. Let $\mathcal{M}$ be the set of all submodules of the module X that are direct sums of modules isomorphic to a right ideal of the ring A. The set $\mathcal{M}$ is nonempty by 6.2.3(f). There exists a partial order in $\mathcal{M}$ such that for any $M, M' \in \mathcal{M}$, the relation $M \subsetneqq M'$ is equivalent to the property that $M' = M \oplus N$ for some $N \in \mathcal{M}$. By the Zorn lemma, the set $\mathcal{M}$ contains at least one maximal element K.

We assume that K is not an essential submodule of the module X. Then there exists a nonzero submodule L of the nonsingular module X with $K \cap L = 0$. By 6.2.3(f), there exists a nonzero right ideal B of the ring A such that the module B_A is isomorphic to some submodule L' of the module L. This contradicts the property that K is a maximal element of the set $\mathcal{M}$.

h. Since the ring A is right finite-dimensional, every nonzero right ideal of the ring A is an essential extension of the finite direct sum of nonzero uniform right ideals. Now the assertion follows from **g**.

i. The assertion follows from **f** and **h**. □

9.1.10 Theorem. Let A be a ring and $G = G(A_A)$. The following conditions are equivalent.

1) Every nonsingular right A-module X that is injective with respect to some essential right ideal of the ring A, is an injective module.

2) Every quasi-injective nonsingular right A-module X that is injective with respect to some essential right ideal of the ring A, is an injective module.

3) Every quasi-injective nonsingular right A-module is an injective module.

4) A/G is a right strongly semiprime ring.

Proof. The implication 1) ⇒ 2) is obvious.

The implication 2) ⇒ 3) follows from 9.1.5(d).

The equivalence of 3) and 4) follows from Theorem 7.2.17.

4) ⇒ 1). Let R be one of the rings A, A/G and M is a right R-module. By 6.1.1(e), the module M is injective if and only if M is injective with respect to the module R_R.

Let $h\colon A \to A/G$ be the natural ring epimorphism and let X be a nonsingular right A-module that is injective with respect to some essential right ideal B of the ring A. By 9.1.9(d), $XG = 0$ and X is a natural nonsingular $h(A)$-module. By 9.1.9(a), $h(B)$ is an essential right ideal of the ring $h(A)$. By 9.1.9(d), the module X is injective with respect to $h(B)$. By Theorem 9.1.6, X is an injective $h(A)$-module. By 9.1.9(e), X is an injective A-module. □

9.1.11. Let A be a ring with right Goldie radical $G(A_A)$ and M is an automorphism-invariant nonsingular right A-module which is an essential extension of direct sum of uniform modules.

a. M is an essential extension of some quasi-injective nonsingular module K that is the direct sum of uniform modules, closed in M.

b. If the factor ring $A/G(A_A)$ is a right strongly semiprime ring, then M is an injective module.

Proof. **a.** By 9.1.2(a), $M = X \oplus Y$, where X is a quasi-injective module, Y is an automorphism-invariant square-free module. Therefore, we can assume that M is an automorphism-invariant square-free module. Since M is an essential extension of the direct sum of uniform submodules, M is an essential extension of some module K which is the direct sum of uniform closed submodules K_i in M, $i \in I$. By 9.1.2(a), K is an automorphism-invariant module. By 6.1.13(b), K is a quasi-injective module.

b. By **a**, M is an essential extension of some quasi-injective nonsingular module K. By Theorem 7.2.17, K is an injective essential submodule of the module M. Therefore, K is an essential direct summand of the module M. Then $M = K$ and M is an injective module. □

9.1.12. Let A be a ring with right Goldie radical $G(A_A)$ and M is an automorphism-invariant nonsingular right A-module. If the factor ring $A/G(A_A)$ is a semiprime right Goldie ring, then M is an injective module.

Proof. By 9.1.9(i), M is an essential extension of direct sum of uniform modules. In addition, the semiprime right Goldie ring $A/G(A_A)$ is a right strongly semiprime ring [86]. By 9.1.11(b), M is an injective module. □

9.1.13 Theorem ([189]). For a ring A with right Goldie radical $G(A_A)$, the following conditions are equivalent.

1) $A/G(A_A)$ is a semiprime right Goldie ring.

2) Any direct sum of automorphism-invariant nonsingular right A-modules is an automorphism-invariant module.

3) Any direct sum of automorphism-invariant nonsingular right A-modules is an injective module.

Proof. The implications 3) ⇒ 2) ⇒ 1) are obvious.

1) ⇒ 3). Let M be a direct sum of automorphism-invariant nonsingular right A-modules M_i, $i \in I$. By 9.1.11, all modules M_i are injective. By Theorem 9.1.10, M is a quasi-injective module. By Theorem 7.2.17, M is an injective module. □

9.1.14 Corollary ([116, Theorem 18]). If A is a semiprime right Goldie ring, then all automorphism-invariant nonsingular right A-modules are injective.

9.2 Automorphism-invariant rings

9.2.1. Let A be a right nonsingular ring with maximal right ring of fractions Q.

a. Q is a right injective regular ring, Q can be naturally identified with the ring End Q_A, and Q_A is the injective hull of the module A_A.

b. The ring A is right automorphism-invariant if and only if A contains all invertible elements of the ring Q.

c. If all closed right ideals of the ring A are ideals, then A is a reduced ring and Q is a right and left injective, strongly regular ring.

Proof. **a.** The assertion is well known; e.g., see [141, Section 3.3.].

b. The assertion follows from **a.**

c. Let $x \in A$, $x^2 = 0$ and let B be a $\cap$-complement of the right ideal xA in A_A. Then $B \cap xA = 0$, $B \oplus xA$ is an essential right ideal, and B is a closed right ideal. By assumption, B is an ideal. Therefore, $xB \subseteq B \cap xA = 0$ and $x(B \oplus xA) = 0$. Then $x = 0$, since the ring A is right nonsingular. Then A is a reduced ring such that all closed right ideals of A are ideals. Consequently, Q is a right and left injective, strongly regular ring [151, Chapter 12, 5.2–5.4.]. □

9.2.2. If A is a right automorphism-invariant, right nonsingular ring, then $A = S \times T$, where S is a right injective regular ring and T is a strongly regular ring that contains all invertible elements of its maximal right ring of fractions.

Proof. In [60, Theorem 7, Theorem 8, Example 9], it is proven that if A is a right nonsingular, right automorphism-invariant ring, then $A = S \times T$, where the ring S is right injective, T_T is a square-free module, any sum of closed right ideals of the ring T is a two-sided ideal that is an automorphism-invariant right T-module, and for any prime ideal P of the ring T that is not essential in T_T, the factor ring T/P is a division ring. By 9.2.1(a), $A = S \times T$, where S is a right injective regular ring, T is a right automorphism-invariant, right nonsingular ring, and every closed right ideal of the ring T is an ideal. By 9.2.1(c), T is a reduced ring. Let Q be the maximal right ring of fractions of the ring T. By 9.2.1(c), T is a reduced ring and Q is a right and left injective, strongly regular ring. To prove that T is a strongly regular ring, it is sufficient to prove that an arbitrary element t of the ring T is the product of a central idempotent by an invertible element. Since t is an element of the strongly regular ring Q, we have $t = eu$, where e is a central idempotent of the ring Q and u is an invertible element of the ring Q. By 9.2.1(b), T contains all invertible elements of the ring Q. Therefore, $u \in T$. Then $e = tu^{-1} \in T$ and every element of the ring T is a product of a central idempotent by an invertible element. □

9.2.3 Theorem ([186]). For a ring A, the following conditions are equivalent.

1) A is a right automorphism-invariant, right nonsingular ring.

2) A is a right automorphism-invariant regular ring.

3) $A = S \times T$, where S is a right injective regular ring and T is strongly regular ring which contains all invertible elements of its maximal right ring of fractions.

Proof. The implication 1) $\Rightarrow$ 3) follows from 9.2.2.

The implication 3) $\Rightarrow$ 2) follows from the property that the direct product of the regular rings S and T is a regular ring.

The implication 2) $\Rightarrow$ 1) follows from the property that every regular ring is right and left nonsingular. □

9.2.4 Corollary ([186]). If A is a right automorphism-invariant, right nonsingular indecomposable ring, then A is a right injective ring.

Corollary 9.2.4 follows from Theorem 9.2.3 and the property that every strongly regular indecomposable ring is a division ring; consequently, it is a right injective ring.

9.2.5 Corollary ([186]). Let A be a right automorphism-invariant, right nonsingular ring that does not contain an infinite set of nonzero central orthogonal idempotents. Then A is a right injective ring.

Corollary 9.2.5 follows from Corollary 9.2.4 and the property that every ring which does not contain an infinite set of nonzero central orthogonal idempotents, is the finite direct product of indecomposable rings.

9.2.6 The completion of the proof of Theorems 9A–9D. Theorems 9A–9D follow from Theorems 9.1.6, 9.1.10, 9.1.13 and 9.2.3.

9.2.7 Open question. Describe right automorphism-invariant group rings.

9.2.8 Open question. Describe right automorphism-extendable (resp., endomorphism-extendable) group rings.

Bibliography

[1] Abbasi A, Hassanzadeh-lelekaami D. Quasi-prime Submodules and Developed Zariski Topology. Algebra Colloquium., 2012, 19(spec01), 1089–1108.

[2] Abuhlail J, Jarrar M, Kabbaj S. Commutative rings in which every finitely generated ideal is quasi-projective. J. Pure Appl. Algebra., 2011, 215, 2504–2511.

[3] Achkar H. Sur les anneaux arithmétiques à gauche. C. R. Acad. Sci., 1974, 278(5), A307–A309.

[4] Achkar H. Sur les propriétés des anneaux arithmétiques à gauche. C. R. Acad. Sci., 1977, 284(17), A993–A995.

[5] Achkar H. Anneaux arithmétiques à gauche. C. R. Acad. Sci., 1978, 286(20), A871–A873.

[6] Achkar H. Anneaux arithmétiques à gauche: propriétés de transfert. C. R. Acad. Sci., 1979, 288(11), A583–A586.

[7] Alahmadi A, Er N, Jain SK. Modules which are invariant under monomorphisms of their injective hulls. J. Australian Math. Soc., 2005, 79(3), 2265–2271.

[8] Albu T, Năstăsescu C. Modules arithmétiques. Acta. Math. Acad. Sci. Hung., 1974, 25(3–4), 299–311.

[9] Amitsur SA. Remarks on principal ideal rings. Osaka Math. J., 1963, 15(1), 59–69.

[10] Anderson DD. Some remarks on the ring $R(X)$. Comment. Math. Univ. St. Pauli., 1977, 26(2), 137–140.

[11] Anderson DD, Anderson DF, Markanda R. The rings $R(X)$ and $R\langle X\rangle$. J. Algebra., 1985, 95(1), 96–115.

[12] Anderson DD, Camillo V. Armendariz rings and Gaussian rings. Comm. Algebra., 1998, 26(7), 2265–2272.

[13] Anderson DD, Chun S. McCoy modules and related modules over commutative rings. Comm. Algebra., 2017, 45(6), 2593–2601.

[14] Anderson DD, Izumi S, Ohno Y, Ozaki M. GCD and LCM-like identities for ideals in commutative rings. J. Algebra Appl., 2016, 15(1), 1650010.

[15] Anderson DD, Markanda R. Unique factorization rings with zero divisors. Houston J. Math., 1985, 11(1), 15–30.

[16] Anderson F, Fuller KR. Rings and Categories of Modules. Springer, Berlin e.a., 1973.

[17] Anh PN, Siddoway M. Divisibility theory of semi-hereditary rings. Proc. Amer. Math. Soc., 2010, 138, 4231–4242.

[18] Anh PN, Siddoway M. Divisibility Theory of Arithmetical Rings with One Minimal Prime Ideal. Comm. Algebra., 2016, 44(2), 823–836.

[19] Atani SE. Strongly irreducible submodules. Bull. Korean Math. Soc., 2005, 42(1), 121–131.

[20] Aubert KE, Beck I. Chinese rings. J. Pure Appl. Algebra., 1982, 24, 221–226.

[21] Azizi A. Strongly irreducible ideals. J.Australian Math. Soc., 2008, 84(2), 145–154.

[22] Azizi A. Radical formula and weakly prime submodules. Glasgow Math. J., 2009, 51, 405–412.

[23] Azizi A. When radical of primary submodules. The Rocky Mountain J. Math., 2011, 41(4), 1045–1058.

[24] Bakkari C, Mahdou N, Mouanis H. Prüfer–like conditions in subring retracts and applications. Comm. Algebra., 2009, 37(1), 47–55.

[25] Bakkari C, Mahdou N. Prüfer–like conditions in pullbacks. In: Fontana M. et al. (eds.). Commutative Algebra and Applications. Walter de Gruyter, 2009, 41–47.

[26] Bandelt H-J, Petrich M. Subdirect products of rings and distributive lattices. , 1982, 25(2), 155–171.

[27] Bazzoni S, Glaz S. Prüfer rings. In: Brewer JW, Glaz S, Heinzer WJ, Olberding BM. (eds.). Multiplicative Ideal Theory in Commutative Algebra. Springer, 2006, 55–72.

https://doi.org/10.1515/9783110659825-010

[28] Bazzoni S, Glaz S. Gaussian properties of total rings of quotients. J. Algebra., 2007, 310, 180–193.

[29] Behrens EA. Distributive Darstellbare Ringe I. Math. Z., 1960, 73(5), 409–432.

[30] Behrens EA. Distributive Darstellbare Ringe II. Math.Z., 1961, 76(4), 367–384.

[31] Behrens EA. Die Halbgruppe der Ideale in Algebren mit distributivem Idealverband. Arch. Math., 1962, 13, 251–266.

[32] Behrens EA. Ring Theory. Acad. Press, New York, 1972.

[33] Behrens EA. Noncommutative arithmetic rings. Rings, Modules and Radicals. North-Holland Pub. Co., Amsterdam–London, 1973, 67–71.

[34] Belzner T. Towards self–duality of semidistributive Artinian rings. J. Algebra., 1990, 135(1), 74–95.

[35] Bessenrodt K, Brungs HH, Törner G. Right chain rings, Part 1. Schriftenreihe des Fachbereich Mathematik, Universität Duisburg, 1990.

[36] Blair RL. Ideal lattice and the structure of rings. Trans. Amer. Math. Soc., 1953, 75(1), 136–153.

[37] Boisen MB, Sheldon PB. The valuation structure of homomorphic images of Prüfer domains. Proc. Amer. Math. Soc., 1974, 46, 335–342.

[38] Boynton J. Pullbacks of Arithmetical Rings. Comm. Algebra., 2007, 35(9), 2671–2684.

[39] Boynton J. Prüfer conditions and the total quotient ring. Comm. Algebra., 2011, 39(5), 1624–1630.

[40] Burgess WD, Stephenson W. An analogue of the Pierce sheaf for non–commutative rings. Comm. Algebra., 1978, 6(9), 863–886

[41] Burgess WD, Stephenson W. Rings all of whose Pierce stalks are local. Canad. Math. Bull., 1979, 22(2), 159–164.

[42] Caballero CE. Self–duality and ℓ–hereditary semidistributive rings. Comm. Algebra., 1986, 14(10), 1821–1843.

[43] Camillo V. Distributive modules. J. Algebra., 1975, 36(1), 16–25.

[44] Chatters AW. A decomposition theorem for Noetherian hereditary rings. Bull. London Math. Soc., 1972, 4, 125–126.

[45] Chapman ST. (Ed.). Arithmetical Properties of Commutative Rings and Monoids. CRC Press, 2005, 416 pp.

[46] Chouinard LG. Transfer maps. Comm. Algebra., 1980, 8(16), 1519–1537.

[47] Chouinard LG, Hardy BR, Shores TS. Arithmetical semihereditary semigroup rings. Comm. Algebra, 1980, 8(17), 1593–1652.

[48] Couchot F. The λ-Dimension of Commutative Arithmetic Rings. Comm. Algebra., 2003, 31(7), 3143–3158.

[49] Couchot F. Modules with RD–Composition Series over a Commutative Ring. Comm. Algebra., 2003, 31(7), 3171–3194.

[50] Couchot F. Localization of Injective Modules Over Arithmetical Rings. Comm. Algebra., 2009, 37(10), 3418–3423.

[51] Couchot F. Almost clean rings and arithmetical rings. In: Fontana, Marco et al. (eds.). Commutative algebra and its applications. Proceedings of the fifth international Fez conference on commutative algebra and applications, Fez, Morocco, June 23–28, 2009. Walter de Gruyter, Berlin, 2009, 135–154.

[52] Couchot F. Finitistic weak dimension of commutative arithmetical rings. Arab. J. Math., 2012, 1(1), 63–67.

[53] Couchot F. Gaussian trivial ring extensions and FQP–rings. Comm. Algebra., 2015, 43(7), 2863–2874.

[54] Couchot F. Weak Dimension of FP-injective Modules Over Chain Rings. Comm. Algebra., 2016, 44(1), 381–389.

[55] Darani AY. Notes on annihilator conditions in modules over commutative rings. An. St. Univ. Ovidius Constanta., 2010, 18(2), 59–72
[56] Dickson S.E, Fuller KR. Algebras for which every indecomposable right module is invariant in its injectve envelope. Pacific J. Math., 1969, 31(3), 655–658.
[57] Dung NV, Huynh DV, Smith PF, Wisbauer R. Extending Modules. John Wiley and Sons Inc., New York., 1994.
[58] Eisenbud D, Griffith Ph. Serial rings. J. Algebra., 1971, 17, 389–400.
[59] Er N. Rings whose modules are direct sums of extending modules. Proc. Amer. Math. Soc., 2009, 137(7), 2265–2271.
[60] Er N, Singh S, Srivastava AK. Rings and modules which are stable under automorphisms of their injective hulls. J. Algebra., 2013, 379, 223–229.
[61] Esin S, Kanuni M, Koc A, Radler K, Rangaswamy KM. On Prüfer-like Properties of Leavitt Path Algebras. 2018, arXiv: 1808.10194.
[62] Faith C. Algebra: Rings, Modules and Categories I. Springer, Berlin–New York., 1973.
[63] Faith C. Algebra II. Springer, Berlin–New York., 1976.
[64] Finn RT, Martinez J, McGovern WW. Commutative Singular f-Rings. In: Ordered Algebraic Structures. Springer, Dordrecht, 1997, 149–166.
[65] Fuchs L. Über die Ideale arithmetischer Ringe. Comment. Math. Helvet., 1949, 23(1), 334–341.
[66] Fuchs L, Heinzer W, Olberding B. Maximal prime divisors in arithmetical rings. In: Chapman ST. (Ed.). Arithmetical Properties of Commutative Rings and Monoids. CRC Press, 2003, 189–204.
[67] Fuchs L, Heinzer W, Olberding B. Commutative ideal theory without finiteness conditions: Primal ideals. Trans. Amer. Math. Soc., 2005, 357, 2771–2798.
[68] Fuchs L, Heinzer W, Olberding B. Commutative ideal theory without finiteness conditions: Completely irreducible ideals. Trans. Amer. Math. Soc., 2006, 358(7), 3113–3131.
[69] Gillman L, Henriksen M. Rings of continuous functions in which every finitely generated ideal is principal. Trans. Amer. Math. Soc., 1956, 82(2), 366–391.
[70] Gillman L, Henriksen M. Some remarks about elementary divisor rings. Trans. Amer. Math. Soc., 1956, 82(2), 362–365.
[71] Gilmer R. A Note on Finitely Generated Ideals of an Arithmetical Ring. In: Badawi A. (Ed.). Trends in Commutative Rings Research. Intern. J., 2003, 27–30.
[72] Glaz S. Prüfer Conditions in Rings with Zero–Divisors. Lecture Notes in Pure and Applied Mathematics., 2005, 241, 272–281.
[73] Glaz S. Finiteness and Homological Conditions in Commutative Group rings. In: Francisco C, Klingler LC, Sather–Wagstaff SM, Vassilev JC. (eds.). Progress in Commutative Algebra 2. Walter de Gruyter, 2012.
[74] Glaz S, Schwarz R. Prüfer conditions in commutative rings. Arab. J. Sci. Engin., 2011, 36, 967–983.
[75] Golod ES. Arithmetical rings, bi–endomorphisms and Gröbner bases. Russian Mathematical Surveys., 2005, 60(1), 163–164.
[76] Golod ES. Distributivity, binary relations, and standard bases. J. Math. Sci. (New York), 2011, 177(6), 862–867.
[77] Golod ES. A remark on commutative arithmetical rings. J. Math. Sci. (New York), 2016, 213(2), 143–144.
[78] Golod ES, Tuganbaev AA. Annihilators and finitely generated modules. J. Math. Sci. (New York), 2018, 233(5), 656–658.
[79] Goodearl KR. Ring Theory. Marcel Dekker, New York., 1976.
[80] Goodearl KR. Von Neumann Regular Rings. Pitman, London., 1979.

[81] Goodearl KR, Warfield RB. An Introduction to Noncommutative Noetherian Rings. Cambridge University Press, Cambridge., 1989.
[82] Gordon R, Robson JC. Krull dimension. Mem. Amer. Math. Soc., 1973, no. 133, 1–78.
[83] Gräter G. General Lattice Theory. Second edition. Burkhäuser, Basel–Boston–Berlin, 2003.
[84] Gräter J. Über die Distributivität des Idealverbandes eines kommutativen Ringes. Monatsh. Math., 1985, 99(4), 267–278.
[85] Guil Asensio PA, Srivastava AK. Automorphism-invariant modules satisfy the exchange property. J. Algebra., 2013, 388, 101–106.
[86] Handelman D. Strongly semiprime rings. Pacific J. Math., 1975, 60(1), 115–122.
[87] Handelman D, Lawrence J. Strongly prime rings. Trans. Amer. Math. Soc., 1975, 211, 209–223.
[88] Hardy BR, Shores TS. Arithmetical semigroup rings. Canad. J. Math., 1980, 32(6), 1361–1371.
[89] Hattori A. A foundation of torsion theory for modules over general rings. Nagoya Math. J., 1960, 17, 147–158.
[90] Heinzer WJ, Ratliff LJ, Rush DE. Strongly irreducible ideals of a commutative ring. J. Pure Appl. Algebra., 2002, 166(3), 267–275.
[91] Heinzer W, Olberding B. Unique irredundant intersections of completely irreducible ideals. J. Algebra., 2005, 287(2), 432–448.
[92] Henriksen M. Some remarks about elementary divisor rings, II. Michigan Math. J., 1956, 3, 159–163.
[93] Hinohara Y. Note on noncommutative local rings. Nagoya Math. J., 1960, 17, 161–166.
[94] Iberkleid W, McGovern W. Wm. A generalization of the Jaffard–Ohm–Kaplansky Theorem. Algebra universalis., 2009, 61(2), 201–212.
[95] Ivanov G, Vamos P. A Characterization of FGC Rings. Rocky Mountain J. Math., 2002, 32(4), 1485–1492.
[96] Jain SK, Singh S. Quasi-injective and pseudo-injective modules. Canadian Math. Bull., 1975, 18(3), 359–366.
[97] Jain SK, Srivastava AK, Tuganbaev AA. Cyclic Modules and the Structure of Rings. Oxford University Press, Oxford, 2012, ISBN: 978-0-19-966451-1, 232 pp.
[98] Jensen CU. A remark on arithmetical rings. Proc. Amer. Math. Soc., 1964, 15(6), 951–954.
[99] Jensen CU. Arithmetical rings. Acta Math. Acad. Sci. Hungar., 1966, 17(1–2), 115–123.
[100] Jeremy L. Modules et anneaux quasi-continus. Canad. Math. Bull., 1974, 17(2), 217–228.
[101] Jespers E. On geometrical Ω-Krull rings. Comm. Algebra., 1983, 11(7), 771–792.
[102] Johnson RE, Wong FT. Quasi-injective modules and irreducible rings. J. London Math. Soc., 1961, 36, 260–268.
[103] Jondrup S. p.p. rings and finitely generated flat ideals. Trans. Amer. Math. Soc., 1971, 28(2), 431–435.
[104] Kabbaj S, Mahdou N, Moutui MA. S. Bi–amalgamations subject to the arithmetical property. J. Algebra Appl., 2017, 16(02), 1750030.
[105] Kabbour M, Mahdou N. Arithmetical property in amalgamated algebras along an ideal. Palestine J. Math., 2014, 3, 395–399
[106] Kaplansky I. Elementary divisors and modules. Trans. Amer. Math. Soc., 1949, 19(2), 21–23.
[107] Khaksari A, Ershad M, Sharif H. Strongly Irreducible Submodules of Modules. Acta Math. Sinica., 2006, 22(4), 1189–1196.
[108] Khashan HA, Al-Ezeh H. Conditions under which $R(x)$ and $R\langle x\rangle$ are almost Q-rings. Arch. Math (Brno)., 2007, 43, 231–236.
[109] Kirichenko VV, Yaremenko YV. On semiperfect semidistributive rings. Matem. Zametki., 2001, 69(1), 153–156.
[110] Koike K. Self–duality of quasi–Harada rings and locally distributive rings. J. Algebra., 2006, 302(2), 613–645.

[111] Kutami M, Oshiro K. Strongly semiprime rings and non-singular quasi-injective modules. Osaka J. Math., 1980, 17, 41–50.
[112] Lam TY. Exercises in Classical Ring Theory. Springer, New York., 1995.
[113] Lam TY. Lectures on Modules and Rings. New York, Springer, 1999.
[114] Larsen MD, Lewis WJ, Shores TS. Elementary divisor rings and finitely presented modules. Trans. Amer. Math. Soc., 1974, 187(1), 231–248.
[115] Lawrence J. A singular primitive ring. Trans. Amer. Math. Soc., 1974, 45(1), 59–62.
[116] Lee TK, Zhou Y. Modules which are invariant under automorphisms of their injective hulls. J. Algebra Appl., 2013, 6(2), 1250159, 9 pp.
[117] Lemonnier B. Sur les anneaux qui ont une déviation. C. R. Acad. Sc. Paris. Ser. A., 1972, T. 275, A357–A359.
[118] Lemonnier B. Dimension de Krull et codéviation des anneaux semi–héréditaires. C. R. Acad. Sc. Paris. Ser. A., 1977, T. 284, A663–A666.
[119] Lenagan TH. Bounded hereditary Noetherian prime rings. J. London Math. Soc., 1973, 6, 241–246.
[120] Levy LS. Sometimes only square matrices can be diagonalized. Proc. Amer. Math. Soc., 1975, 52, 18–22.
[121] Lu X. Some properties of Completely arithmetical rings. Algebra Coll., 2016, 23(1), 83–88.
[122] Lu X, Boynton J. Completely arithmetical rings. Comm. Algebra., 2014, 42(9), 4047–4054.
[123] Lucas TG. Strong Prüfer rings and the ring of finite fractions. J. Pure Appl. Algebra., 1993, 84, 59–71.
[124] Lucas TG. Gaussian polynomials and invertibility. Proc. Amer. Math. Soc., 2005, 133(7), 1881–1886.
[125] Lucas TG. The Gaussian property for rings and polynomials. Houston J. Math., 2008, 34(1), 1–18.
[126] McGovern W, Sharma M. Gaussian property of the rings $R(X)$ and $R\langle X\rangle$. Comm. Algebra., 2016, 44(4), 1636–1646.
[127] Menzel W. Ein Kriterium für Distributivität des Untergruppenverbands einer Abelschen Operatorgruppe. Math. Z., 1961, 75(3), 271–276.
[128] Michler G, Wille R. Die primitiven Klassen arithmetischer Ringe. Math. Z., 1970, 113(5), 369–372.
[129] Min LS. On the constructions of local and arithmetical rings. Acta. Math. Acad. Sci. Hung., 1978, 32(1–2), 31–34.
[130] Mishina AP. On automorphisms and endomorphisms of Abelian groups [In Russian]. Moscow Univ. Math. Bull., 1972, no. 1, 62–66.
[131] Mohamed SH, Müller BJ. Continuous and Discrete Modules. Cambridge University Press, Cambridge., 1990.
[132] Nicholson WK. Lifting idempotents and exchange rings. Trans. Amer. Math. Soc., 1977, 229(2), 269–278.
[133] Nikseresht A, Aziz A. On Arithmetical Rings and the Radical Formula. Vietnam J. Math., 2010, 38(1), 55–62.
[134] Nikseresht A, Aziz A. On Radical Formula in Modules. Glasgow Math. J., 2011, 53(3), 657–668.
[135] Oystaeyen F, van. Generalized Rees rings and arithmetical graded rings. J. Algebra., 1983, 82(1), 185–193.
[136] Parkash A. Arithmetical rings satisfy the radical formula. J. Comm. Algebra., 2012, 4(2), 293–296.
[137] Puninski GE. Projective modules over the endomorphism ring of a biuniform module. J. Pure Appl. Algebra, 2004, 188(1–3), 227–246.

[138] Rajaee S. Multiplication Modules on Arithmetical Rings. Intern. J. Algebra., 2013, 7(17), 825–828.
[139] Rangaswamy KM. The multiplicative ideal theory of Leavitt path algebras. J. Algebra., 2017, 487, 173–199.
[140] Repnickii VB. Arithmetical Modules. Sbornik: Mathematics., 1981, 38(1), 143–149.
[141] Rowen LH. Ring Theory. Volume I. Academic Press, Boston, 1988.
[142] Schwarz R, Glaz S. Commutative group rings with von Neumann regular total rings of quotients. J. Algebra., 2013, 388, 287–293.
[143] Shores T, Lewis WJ. Serial modules and endomorphism rings. Duke Math. J., 1974, 41(4), 889–909.
[144] Singh S. On pseudo-injective modules. Riv. mat. Univ. Parma., 1969, 9, 59–65.
[145] Singh S. Quasi-injective and quasi-projective modules over hereditary Noetherian prime rings. Canad. J. Math., 1974, 26(5), 1173–1185.
[146] Singh S. Modules over hereditary Noetherian prime rings. Canad. J. Math., 1975, 27(4), 867–883.
[147] Singh S, Jain SK. On pseudo-injective modules and self–pseudo-injective rings. J. Math. Sci., 1967, 2, 23–31.
[148] Singh S, Srivastava AK. Rings of invariant module type and automorphism-invariant modules. Ring theory and its applications. Contemp. Math., Amer. Math. Soc., Providence, RI, 2014, 609, 299–311.
[149] Small LW. Semihereditary rings. Bull. Amer. Math. Soc., 1967, 73(5), 656–658.
[150] Snashal N. Scalar-reflexity and FGC rings. J. Algebra., 1994, 165, 380–388.
[151] Stenström B. Rings of Quotients. Springer, Berlin, New York, 1975.
[152] Stephenson W. Modules whose lattice of submodules is distributive. Proc. London Math. Soc., 1974, 28(2), 291–310.
[153] Tachikawa H. QF–3 rings and categories of projective modules. J. Algebra., 1974, 28(3), 408–413.
[154] Teply ML. Pseudo-injective modules which are not quasi-injective. Proc. Amer. Math. Soc., 1975, 49(2), 305–310.
[155] Tuganbaev AA. Structure of modules close to injective. Siberian Mathematical Journal., 1977, 18(4), 631–637.
[156] Tuganbaev AA. Characterizations of rings using skew-injective and skew-projective modules [In Russian]. Moscow University Mathematics Bulletin., 1979, no. 3, 48–51.
[157] Tuganbaev AA. Distributive modules. Russian Math. Surveys., 1980, 35(5), 275–276.
[158] Tuganbaev AA. Integrally closed rings. Math. USSR–Sb., 1982, 43(4), 485–498.
[159] Tuganbaev AA. Semi-injective modules. Math. Notes., 1982, 31(3), 230–234.
[160] Tuganbaev AA. Small-injective rings. Russian Math. Surveys., 1982, 37(5), 196–197.
[161] Tuganbaev AA. Rings with flat right ideals and distributive rings. Math. Notes., 1985, 38(2), 631–636.
[162] Tuganbaev AA. Rings with skew-injective cyclic modules [In Russian]. Abelian groups and modules. Tomsk State University, Tomsk, 1986, 151–158.
[163] Tuganbaev AA. Rings over which all cyclic modules are skew-injective. J. Soviet Math., 1986, 33(4), 1153–1157.
[164] Tuganbaev AA. Rings with skew-injective factor rings. Soviet Math. (Iz. VUZ)., 1991, 35(1), 97–108.
[165] Tuganbaev AA. Flat modules and rings finitely generated as modules over their centers. Math. Notes., 1996, 60(2), 186–203.
[166] Tuganbaev AA. Semidistributive Modules and Rings. Kluwer Academic Publishers, Dordrecht–Boston–London, 1998.

[167] Tuganbaev AA. Skew-injective modules. J. Math. Sci. (New York), 1999, 57(3), 613–614.
[168] Tuganbaev AA. Distributive non-localizable rings. Russian Math. Surveys., 2002, 57(3), 613–614.
[169] Tuganbaev AA. Structure of distributive rings. Sb. Math., 2002, 193(5), 745–760.
[170] Tuganbaev AA. Rings Close to Regular. Kluwer Academic Publishers, Dordrecht–Boston–London, 2002.
[171] Tuganbaev AA. Multiplication modules. J. Math. Sci. (New York), 2004, 123(2), 3839–3905.
[172] Tuganbaev AA. Distributive and multiplication modules and rings. Math. Notes., 2004, 75(3), 391–400.
[173] Tuganbaev AA. Rings over which all modules are completely integrally closed. Discrete Math. Appl., 2011, 21(4), 477–497.
[174] Tuganbaev AA. Automorphisms of submodules and their extensions. Discrete Math. Appl., 2013, 23(1), 115–124.
[175] Tuganbaev AA. Characteristic submodules of injective modules. Discrete Math. Appl., 2013, 23(2), 203–209.
[176] Tuganbaev AA. Extensions of automorphisms of submodules. J. Math. Sci. (New York), 2015, 206(5), 583–596.
[177] Tuganbaev AA. Automorphism-invariant modules. J. Math. Sci. (New York), 2015, 206(6), 694–698.
[178] Tuganbaev AA. Characteristic submodules of injective modules over strongly prime rings. Discrete Math. Appl., 2014, 24(4), 253–256.
[179] Tuganbaev AA. Modules over strongly prime rings. J. Algebra Appl., 2015, 14(5), 1550076, 9 pp.
[180] Tuganbaev AA. Rings whose finitely generated right ideals are quasi-projective. Discrete Math. Appl., 2015, 25(4), 245–251.
[181] Tuganbaev AA. Automorphism-extendable modules. Discrete Math. Appl., 2015, 25(5), 305–309.
[182] Tuganbaev AA. Arithmetical rings and quasi-projective ideals. J. Math. Sci. (New York), 2016, 213(2), 268–271.
[183] Tuganbaev AA. Bezout rings without non–central idempotents. Discrete Math. Appl., 2016, 26(6), 369–377.
[184] Tuganbaev AA. Automorphism-extendable semi–Artinian modules. J. Algebra Appl., 2017, 16(2), 1750029, 5 pp. doi:10.1142/S0219498817500293.
[185] Tuganbaev AA. Modules over strongly semiprime rings. Discrete Math. Appl., 2019, 29(2), 143–147.
[186] Tuganbaev AA. Automorphism-invariant non-singular rings and modules. J. Algebra., 2017, 485, 247–253.
[187] Tuganbaev AA. Bezout rings, annihilators, and diagonalizability. J. Math. Sci. (New York), 2019, 237(2), 329?331.
[188] Tuganbaev AA. Arithmetical Rings and Krull Dimension. Discrete Math. Appl., 2018, 28(2), 113–118.
[189] Tuganbaev AA. Injective and automorphism-invariant non-singular modules. Comm. Algebra., 2018, 46(4), 1716–1721.
[190] Vamos P. The decomposition of finitely generated modules and fractionally self-injective rings. J. London Math. Soc., 1977, 16, 209–220.
[191] Vechtomov EM. Distributive rings of continuous functions and F–spaces. Math. Notes., 1983, 34(3), 643–648.
[192] Warfield RB. Decomposability of finitely presented modules. Proc. Amer. Math. Soc., 1970, 25, 167–172.

[193] Warfield RB. Stable generation of modules. Lecture Notes Math., 1979, 700, 16–33.

[194] Wisbauer R. Foundations of Module and Ring Theory. Gordon and Breach, Philadelphia., 1991.

[195] Wright MH. Right locally distributive rings. Ring Theory. Proc. Bien. Ohio State; Denison Conf., Granville, Ohio, May 1992, World Scientific, Singapore etc., 1993, 350–357.

[196] Xue W. Rings with Morita duality. Lecture Notes Math., 1992, 1523, 1–197.

[197] Yaremenko Yu. Noetherian semi-perfect rings of distributive module type. Mat. Stud., 1997, 8(1), 3–10.

[198] Yukimoto Y. Artinian rings whose projective indecomposables are distributive. Osaka J. Math., 1985, 22, 339–344.

Index

https://doi.org/10.1515/9783110659825-011

Bei Fragen zur Produktsicherheit wenden Sie sich bitte an:
If you have any questions regarding product safety,
please contact:

Walter de Gruyter GmbH
Genthiner Straße 13
10785 Berlin
productsafety@degruyterbrill.com